SPRINGER TRACTS
IN MODERN PHYSICS

Ergebnisse
der exakten Natur-
wissenschaften

Volume **71**

Editor: G. Höhler

Associate Editor: E. A. Niekisch

Editorial Board: S. Flügge J. Hamilton F. Hund
H. Lehmann G. Leibfried W. Paul

Springer-Verlag Berlin Heidelberg GmbH 1974

Manuscripts for publication should be addressed to:

G. HÖHLER, Institut für Theoretische Kernphysik der Universität, 75 Karlsruhe 1, Postfach 6380

Proofs and all correspondence concerning papers in the process of publication should be addressed to:

E. A. NIEKISCH, Institut für Grenzflächenforschung und Vakuumphysik der Kernforschungsanlage Jülich, 517 Jülich, Postfach 365

ISBN 978-3-662-15869-2   ISBN 978-3-540-37920-1 (eBook)
DOI 10.1007/978-3-540-37920-1

Nuclear Physics

Contents

Study of Nuclear Structure by Muon Capture

H. ÜBERALL

Contents

Introduction

In a talk delivered at the 1966 Williamsburg Conference on Inter-mediate Energy Physics [1], the present author has pointed out that muon capture in nuclei, which until then had only been used for determining the weak coupling constants and ascertaining the universality of the weak interactions, could with much greater advantage be employed as an effective tool for studying nuclear structure. This includes the structure of the ground state of the capturing nucleus and of all the $T_z - 1$ analogs of its excited states, especially of the giant resonances. Subsequent muon capture experiments, carried out at many laboratories, indeed followed this suggestion, as has been described in a series of more recent review articles and talks [2–6]. In the present report, we shall attempt to review recent developments in muon capture as used for the study of nuclear structure, mainly in light nuclei. This will include total capture rates, excitations of giant resonances of both electric and magnetic type, particle and photon emission following muon capture, neutron asymmetries from polarized muons, and Doppler broadening of photon lines due to angular correlations.

Other reactions that are closely related to the muon capture reaction have also been considered as possible tools for nuclear structure studies in the meantime; among those figure, for example [7], the production of charged pions by photons as well as by electrons [8], but most prominently the radiative capture of stopped negative pions. A series of important experiments has been carried out also for this latter process [9]. It goes without saying that the whole field of muon and pion capture studies is expected to experience a big advance with the completion of the three so-called "meson factories" at Los Alamos (LAMPF), Zürich (SIN), and Vancouver (TRIUMF), which is now imminent.

I. Total Muon Capture Rates

Calculations of nuclear muon capture phenomena have usually been carried out in the impulse approximation, and are based on the "Primakoff Hamiltonian" [10], a non-relativistic approximation for the universal Fermi interaction in which some relativistic terms are included, and some meson effects such as form factors and an induced pseudoscalar interaction are taken into account; in addition, the "weak magnetism" term from the conserved vector current hypothesis is considered. This leads to the expression for the total capture rate [11]

$$
\begin{aligned}
\Lambda_c = (m_\mu^2/2\pi)\,\langle|\varphi_\mu|^2\rangle\,[G_V^2 M_V^2 + 3G_A^2 M_A^2 \\
+ (G_P^2 - 2G_P G_A)\,M_P^2] + \Lambda_c'
\end{aligned}
\tag{1}
$$

where m_μ = muon mass, $\langle|\varphi_\mu|^2\rangle$ = average squared $1s$-wave function of the bond muon; G_V, G_A, and G_P are the effective Fermi, Gamow-Teller and induced pseudoscalar coupling constants, respectively of the Primakoff Hamiltonian, and Λ_c' contains the ($\sim 10\%$) corrections to the capture rate due to those nucleon recoil corrections not included in the G's. The squared matrix elements are

$$
M_V^2 = (2J_i + 1)^{-\frac{1}{2}} \sum_{M_i M_f} (v_{if}/m_\mu)^2
$$
$$
\cdot \int (d\Omega_v/4\pi) \left| \left\langle J_f M_f \left| \sum_{j=1}^{A} \tau_j^- \exp(-i\boldsymbol{v}_{if} \cdot \boldsymbol{r}_j) \right| J_i M_i \right\rangle \right|^2 ,
\tag{2a}
$$

$$
M_A^2 = \tfrac{1}{3}(2J_i + 1)^{-\frac{1}{2}} \sum_{M_i M_f} (v_{if}/m_\mu)^2
$$
$$
\cdot \int (d\Omega_v/4\pi) \left| \left\langle J_f M_f \left| \sum_{j=1}^{A} \tau_j^- \boldsymbol{\sigma}_j \exp(-i\boldsymbol{v}_{if} \cdot \boldsymbol{r}_j) \right| J_i M_i \right\rangle \right|^2 ,
\tag{2b}
$$

$$
M_P^2 = (2J_i + 1)^{-\frac{1}{2}} \sum_{M_i M_f} (v_{if}/m_\mu)^2
$$
$$
\cdot \int (d\Omega_v/4\pi) \left| \left\langle J_f M_f \left| \sum_{j=1}^{A} \tau_j^- \hat{v} \cdot \boldsymbol{\sigma}_j \exp(-i\boldsymbol{v}_{if} \cdot \boldsymbol{r}_j) \right| J_i M_i \right\rangle \right|^2 ,
\tag{2c}
$$

where $|J_i M_i\rangle$ and $|J_f M_f\rangle$ describe initial and final nuclear states, and $\boldsymbol{v}_{if} = v_{if}\hat{v}$ is the neutrino momentum corresponding to the nuclear transition $i \to f$, i.e.

$$\mu^- + (^AZ)_i \to {}^A(Z-1)_f + \nu_\mu . \tag{3}$$

In the previous calculations of total muon capture rates, the following different approaches to the evaluation of Eq. (1) have been taken:

a) Use of the Closure Relation, in order to sum over all (accessible and inaccessible) final nuclear states $|J_f M_f\rangle$. The uncertainty of this method consists in the choice of an average neutrino momentum $\bar{\nu}$ which has to be taken out of the closure sum, but the results of such a calculation have correctly reproduced the trend of capture rates as functions of A and $(N-Z)$, leading to the well-known Primakoff formula [10].

b) Shell Model Calculation, based on the assumption of independent, non-interacting nucleons [12]. With a harmonic oscillator shell model, these calculations lead to the relation

$$M_V^2 = M_A^2 = M_P^2 . \tag{4}$$

The theoretical capture rates obtained in this way always exceed the measured ones [13, 14] by at least a factor of two:

$$\begin{aligned}
\Lambda_c^{\text{theor.}} &\sim 2\Lambda_c^{\text{exp.}}(\text{Ca}), \\
&\sim 3.5\,\Lambda_c^{\text{exp.}}(\text{Pb}).
\end{aligned} \tag{5}$$

c) Fermi Liquid Theory or Migdal Theory [15–17], a method that takes into account the strong interaction between nucleons via the introduction of quasiparticles. Calculations of total muon capture rates have been carried out in this theory by Bunatyan [18] and Novikov [19], and led to excellent agreement with the measured rates. The reason for this agreement, as analyzed by Rho [20], is the fact that M_A is renormalized (reduced) by the interaction while M_V is not, thus leading to the inequality

$$M_A^2 < M_V^2 . \tag{6}$$

Parenthetically, we should point out, however, that the same Migdal theory when applied to a calculation of π^+ photoproduction in ^{16}O for the partial transition rates to the 0^-, 1^-, 2^-, and 3^- states in ^{16}N,

$$\gamma + {}^{16}\text{O} \to {}^{16}\text{N}_{0^-, 1^-, 2^-, 3^-} + \pi^+ \tag{7}$$

predicts only $\sim 40\%$ of the experimental cross section [21]. This fact will be considered again below.

d) The Resonance Model. It was argued by Foldy and Walecka [11] that the main contribution to the capture rate should be provided by the (first forbidden) transition to the nuclear giant resonance; this follows from the similarity of the weak vector interaction to the vector part of the electro-magnetic interaction [22], the latter providing the dominant giant dipole resonance transition in photonuclear absorption [23]. It also follows from an analysis of Tolhoek's shell model work [12], which shows $l = 1$ transitions to predominate. This holds

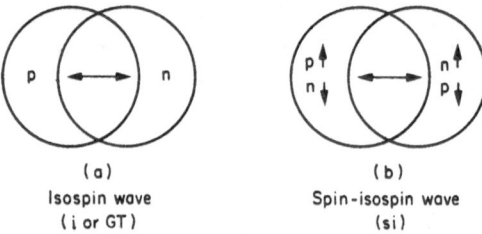

(a)   (b)

Isospin wave   Spin-isospin wave
(i or GT)   (si)

Fig. 1. a Charge-flip (isospin wave) and b Spin-flip (spin-isospin wave) components of the nuclear giant resonance

for both the Fermi and the Gamow-Teller transitions, cf. Eq. (4), and thus implies the existence of a "spin flip giant resonance", clearly confirmed in electroexcitation experiments [7]. Since the nuclear forces are approximately spin independent, one may combine the charge flip and spin flip components of the giant resonance into one 15 dimensional SU_4 supermultiplet [24]. The charge flip components (*i*) are described physically by the Goldhaber-Teller [25] picture of protons oscillating collectively against neutrons (Fig. 1a), and the spin flip components (*si*) by protons with spin-up and neutrons with spin-down oscillating against neutrons with spin-up and protons with spin-down [26], (Fig. 1b). In this simple model, which gives the same transition probabilities as the harmonic-oscillator independent particle shell model [27], the contributions to M_V come exclusively from the *i*-transition, those to M_A (and M_P) from the *si*-transition. The relation of Eq. (4) holds again in the limit of exact SU_4 symmetry.

Foldy and Walecka have obtained M_V^2 by a semiempirical method, in which the "unretarded" part of M_V^2 (at zero momentum transfer) was related to the observed photonuclear cross section $\sigma_\gamma(E)$ of the giant electric dipole resonance,

$$(M_V^2)_{\text{UD}} = \frac{m_\mu^2}{2\pi^2 \alpha} \int\limits_0^{m_\mu} \left(\frac{m_\mu - E}{m_\mu} \right)^4 \frac{\sigma_\gamma(E)}{E} \, dE, \qquad (8)$$

integrated over all photon energies E accessible in muon capture; here $\alpha = 1/137$. Retardation was taken into account by multiplying $(M_V^2)_{UD}$ with the square of the elastic electron scattering form factor at the momentum transfer appropriate to muon capture,

$$q = m_\mu - \bar{\omega}, \tag{9}$$

ω being the nuclear giant resonance energy measured from the ground state of the capturing nucleus. The corresponding values of M_A^2, M_P^2 were then found from Eq. (4). Numerical results were obtained in this way for ^{4}He, ^{12}C, ^{16}O, and ^{40}Ca; they will be quoted below. For ^{16}O and ^{40}Ca, the agreement with experiment was excellent, but for ^{4}He and ^{12}C, the predictions were somewhat too low.

Several studies have been made concerning the effects of SU$_4$ breaking and a corresponding invalidation of Eq. (4). A calculation of Walker [28] attempted to improve the shell model results by admitting realistic (spin-dependent) residual particle-hole interactions for the giant resonance states (described in the particle-hole picture [29]). For ^{16}O, an admixture of 2 particle – 2 hole configurations in the ground state was taken into account, with corresponding 3 particle – 3 hole configurations in the giant resonance. All these effects were found to lead to a reduction of the capture rate to the giant resonances by no more than 10% as compared to the independent particle shell model prediction (which latter exceeded the experimental rate by a factor of two). As to the total rate, no reduction was found at all, due to the appearance of a strong $0^+ \rightarrow 1^+$ transition of $M1$ states predicted in the 20–60 MeV region of excitation by the introduction of the n-particle – n-hole configurations. We shall have to say more about the role of $M1$ transitions in muon capture later on.

An empirical approach to SU$_4$ breaking was taken by Cannata, Leonardi, and Rosa-Clot [30]. This is based on Eq. (8) in which the main contribution to M_V^2 is provided by the giant dipole resonance peak, $E = \omega_i \simeq 22$ MeV (for light nuclei). Empirically, the $0^+ \rightarrow 1^-$ spin flip resonance peaks at a higher energy ω_{si}, however, so that M_A^2 is no longer given by Eqs. (8) and (4). Instead, one finds the values given in Table 1, Column 3. The agreement of the corresponding predicted capture rates (Column 4) with experiment (Column 5) is excellent; the disagreement for ^{12}C can be eliminated by adding to the rate the calculated contribution for the strong $0^+ \rightarrow 1^+$ ($M1$) transition to the ^{12}B ground state (another example of a "giant $M1$" transition mentioned before), which brings the total rate up to 0.36×10^5 sec^{-1}. The calculated rates were obtained with a ratio $g_A/g_V = -1.28$ of Gamow-Teller and Fermi coupling constants for beta decay, as indicated by recent experiments [31]. The earlier predictions of Λ_c by

Table 1. SU$_4$ breaking effects and muon capture rates in some light nuclei

Target	$\omega_{si} - \omega_i$ (MeV)	M_A^2/M_V^2	$\Lambda_c^{\text{theor.}}(\sec^{-1})$	$\Lambda_c^{\text{exp.}}(\sec^{-1})$	$\Lambda_c^{\text{FW}}(\sec^{-1})$
^{4}He	0.9	1	300	336 ± 75	249
^{12}C	2	0.91	0.28×10^5	$(0.36 \pm 0.01) \times 10^5$	0.26×10^5
^{16}O	3	0.86	0.98×10^5	$(0.97 \pm 0.03) \times 10^5$	0.95×10^5
^{40}Ca	7	0.73	25.2×10^5	$(25.5 \pm 0.5) \times 10^5$	28.3×10^5

Foldy and Walecka [11], listed in the last column of Table 1, had been obtained with the older value $g_A/g_V = -1.15$. Using here -1.28 instead will cause Λ_c^{FW} to considerably exceed Λ_c^{exp} except for ^{4}He. One may note that theoretically, the splitting of ω_i and ω_{si} stems from SU$_4$ breaking components in the nucleon-nucleon interaction. Indeed, one may estimate [30] the splitting as

$$\omega_{si} - \omega_i = 2 \langle V_B \rangle, \tag{10}$$

being determined by an average over the Bartlett force.

Cannata [32] has noted that in the photopion reaction, Eq. (7), no reduction of M_A^2 takes place comparable to that given in Column 3 of Table 1; the SU$_4$ results should therefore apply here, while the Migdal theory provides a reduction of M_A^2 and leads to disagreement with experiment, as noted earlier. For the muon capture rates Λ_c, the theories of Migdal and of Cannata provide two alternate approaches which both lead to agreement with experiment, however.

II. Neutron Angular Distribution

In the muon capture reaction, Eq. (3), the final nuclear state $^A(Z-1)_f$ is often excited above its particle threshold, and then decays by emission of one or more neutrons; this sequence is referred to as the "resonance mechanism". In addition, there may be "direct neutron emission" from the nucleus via the basic reaction

$$\mu^- + p \rightarrow n + \nu_\mu, \tag{11}$$

which mainly takes place at excitation energies of and above the giant resonances (≥ 15–20 MeV). Since the muon beam usually has a high degree of polarization, of which $\sim 15\%$ (for spinless nuclei) is retained as the muon reaches the $1s$ orbit, an angular asymmetry of the direct neutrons with respect to the muon spin is expected to result due to the parity non-conservation in muon capture. This asymmetry does not persist in the resonance capture, thus has to be looked for at the high-energy emitted neutrons. For muon capture by a free proton, the asymmetry coefficient α in the general angular distribution

$$\Lambda_\mu \propto 1 + \alpha \cos\vartheta \tag{12}$$

(ϑ being the angle between neutron direction and muon spin) was estimated as $\alpha \simeq -0.4$, being sensitive to the induced pseudoscalar coupling.

The neutron asymmetry problem has constituted a long-standing puzzle in muon capture physics, both experimentally and theoretically. The older experiments [33], carried out for ^{12}C, ^{28}Si, ^{32}S, and ^{40}Ca, find for high-energy neutrons ($E_n > 20$ MeV) very large negative asymmetry coefficients, $\alpha = -0.20$ to -1.0 (largest at the top of the spectrum), which were in definite disagreement with the theoretical results for α obtained using the Primakoff Hamiltonian, the latter being negative also, but no larger than $|\alpha| \lesssim 0.10$. Newer experiments [34] find α positive and moderately large (although with a large degree of uncertainty).

If the latter results are accepted (which really should be done only after a full experimental clarification of the discrepancies), then the asymmetry puzzle seems now to be cleared up by recent calculations [35] that took into account relativistic terms in the capture interaction not contained in the Primakoff Hamiltonian. In the older calculations, small negative values of α were found as mentioned, both in an independent-particle shell model [36], and in a Fermi gas model [37]. More recent shell model calculations of Bogan [38] did not drop the first-order ($\propto 1/m_p$) nucleon-momentum dependent terms as done in the Primakoff Hamiltonian; Bogan also took into account final-state interactions by using an effective neutron momentum p_n' instead of p_n,

$$p_n' = p_n (1 - \bar{V}/2E_n) \tag{13}$$

with $\bar{V}$ an average optical-model potential. This approach yielded small positive values of α.

Piketty's calculation [35] keeps all relativistic terms up to $1/m_p^2$ in a shell-model approach, and is relativistically exact in a Fermi model picture (the latter showing that there was no need to go beyond $1/m_p^2$). While for the Fermi gas, α was found small but still negative, the independent-particle shell model gave α positive and increasing with neutron energy. Figures 2–4 show the shell model results for α vs. E_n for ^{28}Si, ^{32}S, and ^{40}Ca, respectively. Curves I and II include $1/m_p$ and $1/m_p^2$ terms, respectively, while III includes both $1/m_p^2$ terms and final-state interactions. The latter are seen to have no sizable

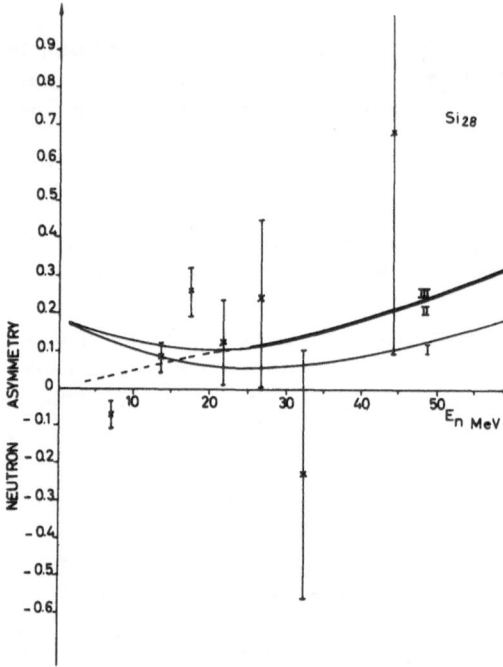

Fig. 2. Neutron asymmetry after muon capture in ^{28}Si; experimental points: Sundelin [34]; theoretical curves: Piketty and Procureur [35]

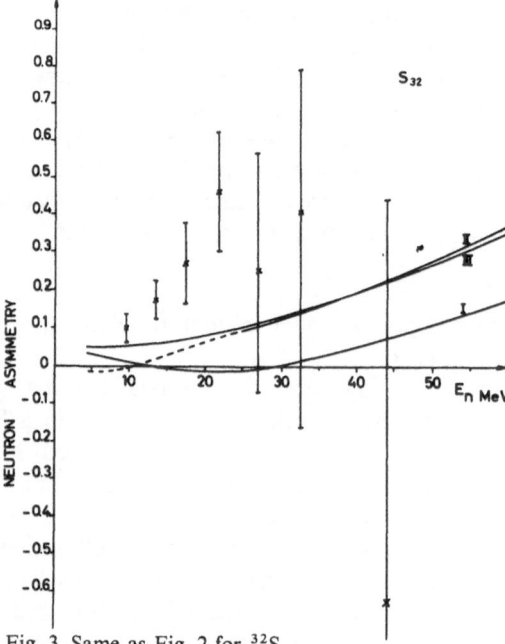

Fig. 3. Same as Fig. 2 for ^{32}S

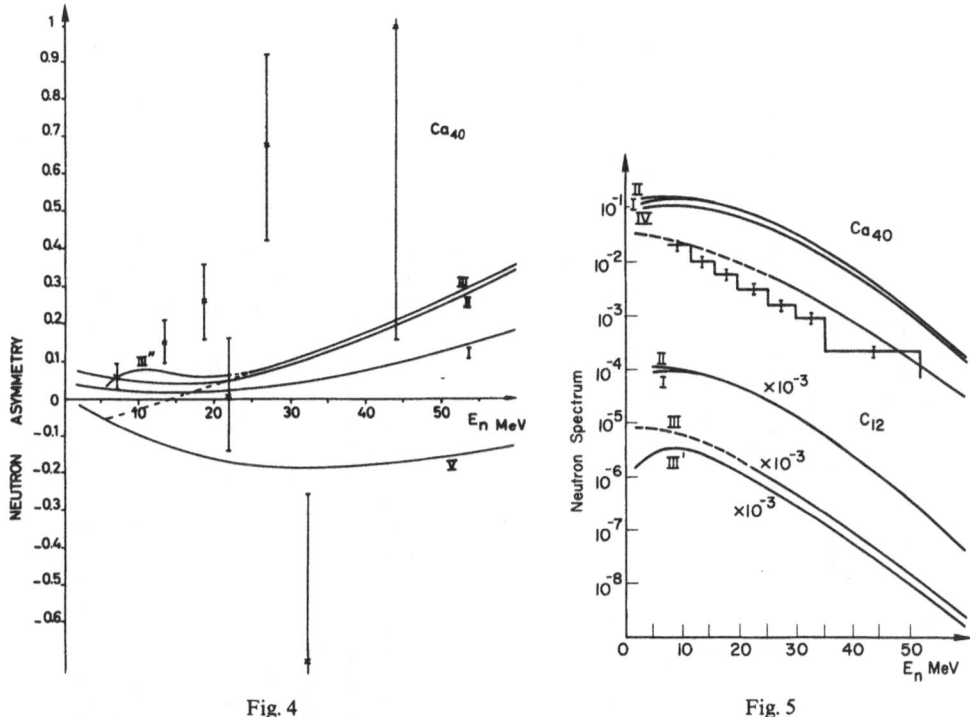

Fig. 4

Fig. 5

Fig. 4. Same as Fig. 2 for ^{40}Ca

Fig. 5. Neutron spectrum after muon capture in ^{40}Ca; experimental curve: Sundelin [34]; theoretical curves: Piketty and Procureur [35]

effect on the asymmetry; they do however greatly suppress the neutron spectrum (Fig. 5) and bring it in agreement with the observed spectrum. The experimental points on Figs. 2–5 are those of Sundelin [34]. For ^{4}He, the calculations of Eramzhian [35] likewise predict large relativistic and final-state effects which render the asymmetry large and positive ($\alpha \lesssim +0.4$).

III. The Giant Resonance Mechanisms

While the direct neutron emission mechanism, Eq. (11), or

$$\mu^- + {}^A Z \rightarrow {}^{A-1}(Z-1) + n + \nu \tag{14}$$

is responsible for the higher-energy neutrons forming the "quasi-elastic continuum", a very important contribution to both total muon capture rates and the spectrum of subsequently emitted neutrons is

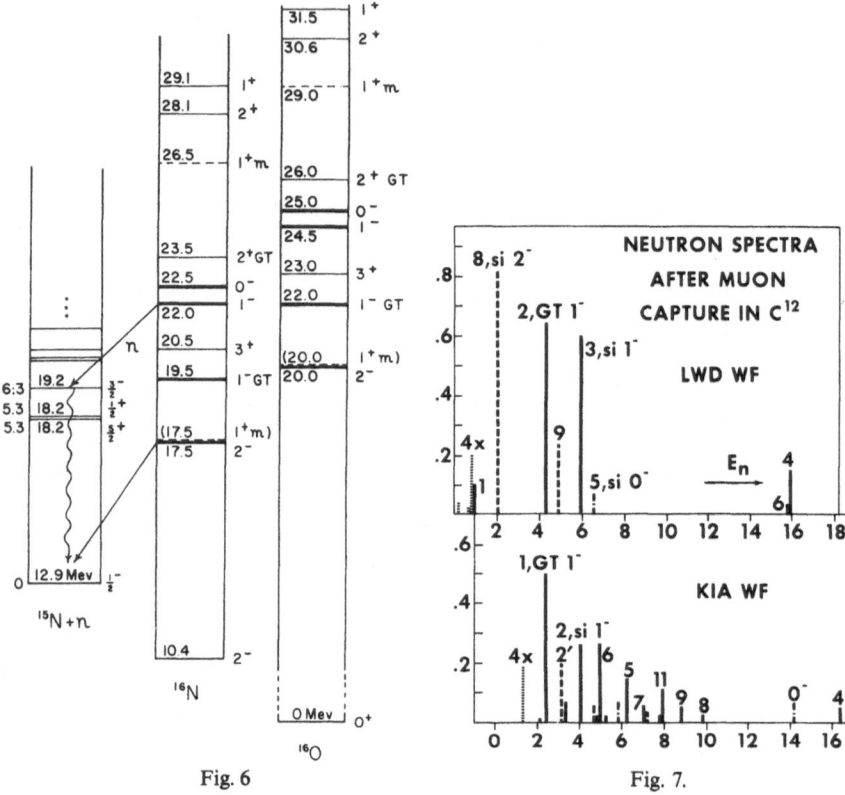

Fig. 6

Fig. 7.

Fig. 6. Spectra of giant resonance states in ^{16}O and ^{16}N and neutron decay channels to ^{15}N

Fig. 7. Calculated [42] neutron spectra after muon capture in ^{12}C. The ^{12}B giant resonances are described by particle-hole models labeled LWD [43], top, and KIA [44], bottom

provided by the resonance mechanism of muon capture [39, 1], mainly via the nuclear giant resonance levels. In this two-step process, the nucleus is first excited to the giant resonance levels as given by Eq. (3) with $f \equiv$ exc. = g.res., and subsequently decays in a variety of ways permitted energetically, e.g.

$$^A(Z-1)_{\text{exc.}} \rightarrow {^A(Z-1)} + \gamma\,, \tag{15a}$$

$$\rightarrow {^{A-1}(Z-1)} + n\,, \tag{15b}$$

$$\rightarrow {^{A-2}(Z-1)} + 2n\,, \tag{15c}$$

$$\rightarrow {^{A-1}(Z-2)} + p\,, \tag{15d}$$

$$\rightarrow {^{A-2}(Z-2)} + n + p\,, \tag{15e}$$

etc., of which the single neutron channel, Eq. (15b), is usually the most prominent decay mode. The daughter nucleus $^{A-A'}(Z-Z')$ may be left in its ground state or in any excited state that is energetically permitted. Parenthetically, we remark that there is really no basic difference between the direct process of Eq. (14) and the resonance process, Eq. (15b); the higher energy resonances become broader, more short-lived, and overlap, thus creating the impression of a nuclear continuum corresponding to direct ejection of a nucleon, Eq. (14). Calculations bear out such a viewpoint [40].

Figure 6 presents a schematic picture of the $T_3 = 0$ giant resonance levels (of dipole and quadrupole type) in ^{16}O and their $T_3 = -1$ analog states in ^{16}N, as well as some possible neutron decay channels to ^{15}N (ground or excited states) [41]. The corresponding appearance of characteristic giant resonance peaks in the spectrum of the emitted neutrons, in exact analogy with the peaks appearing in the particle spectrum in photonuclear (γ, n) and (γ, p) reactions, is a direct indication of the correctness of the resonance mechanism in muon capture [1]. Photon absorption in ^{16}O leads to the resonance states in the same nucleus, which will decay e.g. by proton emission to the final states of ^{15}N shown in Fig. 6. The classification of resonance levels in this figure was obtained by the schematic Goldhaber-Teller model as generalized by the author [26]. A calculation of the neutron spectrum following muon capture in ^{12}C was made [42] using more detailed particle-hole models of the giant resonance due to Lewis [43] and Kamimura [44]. The results are shown in Fig. 7; the peaks are due to neutron decay of the $T_3 - 1$ giant resonance levels in ^{12}B mainly to the ground state of ^{11}B.

The first experiment that showed neutron peaks of this nature after muon capture in ^{32}S was done by Evseyev [45]. His results are shown in Fig. 8; they were obtained by unfolding of the neutron spectrum from the spectrum of knock-on protons observed in a scintillation counter. This author also observed the neutron spectrum from muon capture in ^{16}O, shown in Fig. 9 [46], comparing favorably with the neutron spectrum obtained theoretically from the Goldhaber – Teller model [41], Fig. 10. In that theoretical calculation, the contributions of both dipole and quadrupole giant resonances were obtained. The quadrupole contributions are shown in the upper part of Fig. 10, and are small; they contribute the portion of the total spectrum above the broken line in the lower part of Fig. 10.

In a later experiment by Plett and Sobottka [47], the same spectrum for ^{16}O as in Fig. 10 was measured with better resolution, and a considerably higher background of very low energy neutrons was obtained than in Evseyev's experiment; see Fig. 11. When this

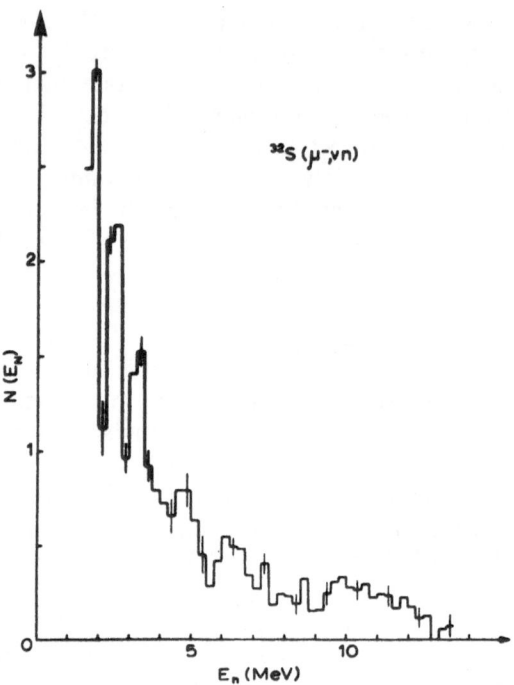

Fig. 8. Experimental neutron spectrum from muon capture in ^{32}S as obtained by Evseyev [45]

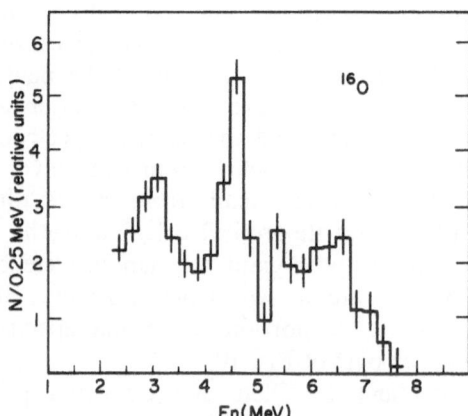

Fig. 9. Experimental neutron spectrum from muon capture in ^{16}O as obtained by Evseyev [46]

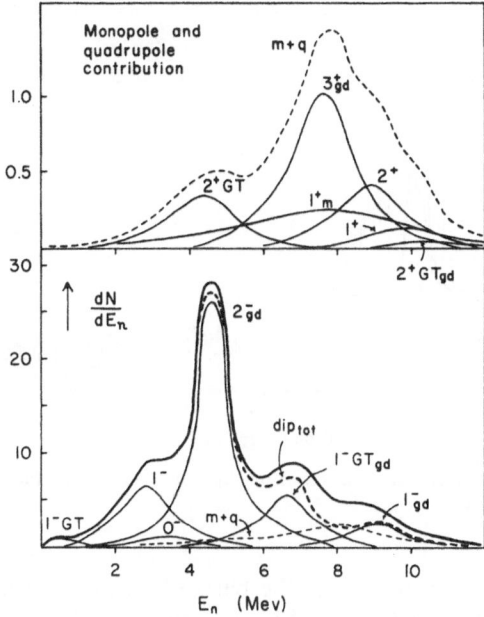

Fig. 10. Neutron spectrum from muon capture in ^{16}O, calculated from the Goldhaber-Teller model [41]. Top portion gives contribution of the quadrupole resonances

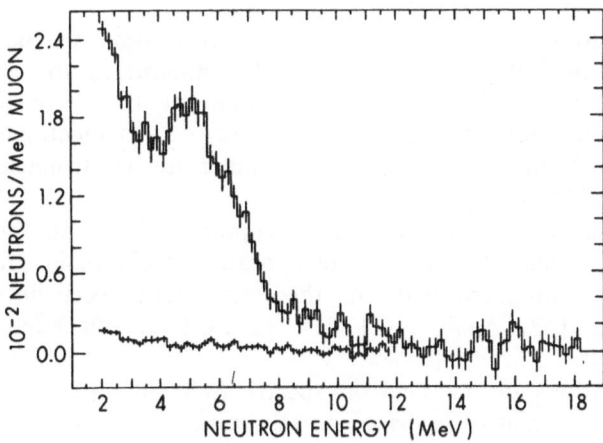

Fig. 11. Experimental neutron spectrum from muon capture in ^{16}O, as obtained by Plett and Sobottka [47]

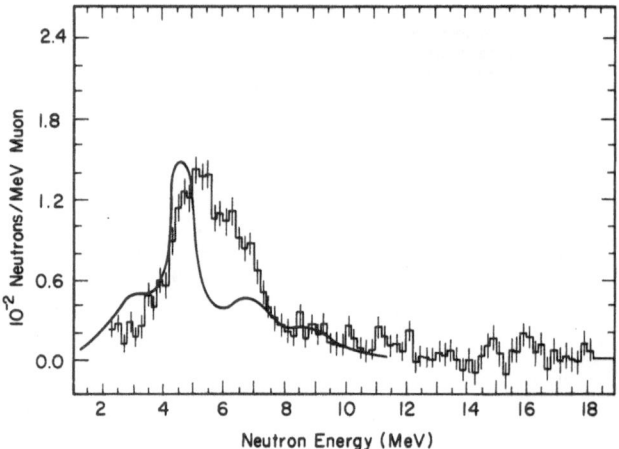

Fig. 12. Same as Fig. 11, with evaporation background subtracted and compared to the theory [41], solid curve

background was subtracted on the basis of an evaporation model, the "giant resonance" spectrum of Fig. 12 resulted, which is compared in this figure with the theoretical spectrum of Reference [41]. The comparison indicates here a shift of ~ 1 MeV between experimental and theoretical levels.

For the 4.6 MeV neutron peak due to the decay of 2^- *si* levels in ^{16}N to the ^{15}N ground state, the experiment finds a 2^- capture-decay rate of $(0.193 \pm 0.55) \times 10^5 \, \mathrm{sec}^{-1}$, while the theory [41] gives $0.417 \times 10^5 \, \mathrm{sec}^{-1}$; the experimental value for the 1^- level in ^{16}N decaying to the ^{15}N ground state (6.6 MeV peak in the neutron spectrum) of $(0.083 \pm 0.033) \times 10^5 \, \mathrm{sec}^{-1}$ compares to the theoretical value of $0.153 \times 10^5 \, \mathrm{sec}^{-1}$. It is seen that the theory overestimates experimental capture rates by a factor of ~ 2, as mentioned earlier for the shell model, and as it is familiar for electronuclear giant resonances [7].

The neutron spectrum after muon capture in ^{12}C, also obtained by Plett [47], is shown in Fig. 13, (evaporation background subtracted), where it is compared with the theoretical KIA spectrum of Kelly and Überall [42]. Again a ~ 1 MeV relative level shift is indicated by these results.

Calculations of the neutron spectra following muon capture in ^{16}O, ^{28}Si, ^{32}S, and ^{40}Ca, based on the particle-hole model, have been carried out by Hill and Überall [48]. In Fig. 14, we show their results for ^{16}O as compared with the experimental results of Plett [47]. Recent measurements of neutron spectra from Ca, Tl, Pb, and Bi were

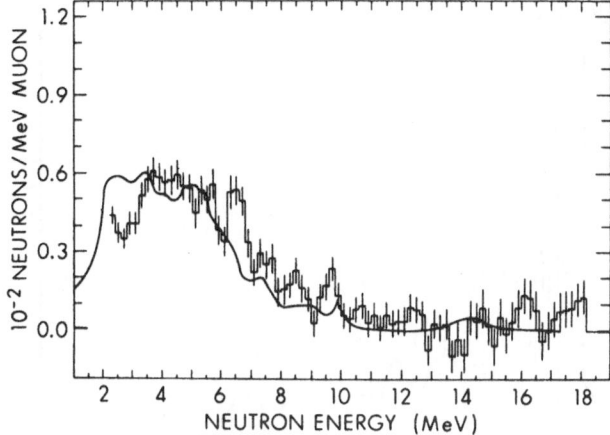

Fig. 13. Experimental neutron spectrum from muon capture in ^{12}C, as obtained by Plett and Sobottka [47]; with theoretical spectrum (KIA) of reference [42]

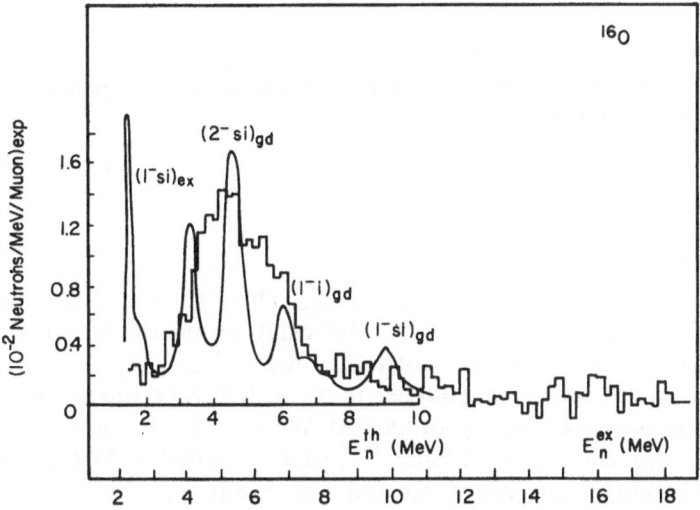

Fig. 14. Neutron spectrum following muon capture in ^{16}O, as calculated on the particle-hole model [48], compared to Plett's [47] experimental data

performed by Schröder et al. [49]. Their results for ^{40}Ca are presented in Fig. 15, and compared to particle-hole theory [48].

As to the emission of protons after muon capture, Eq. (15d), it has been pointed out [50] that this may provide information about the admixture of two particle, two hole states into·the giant resonance.

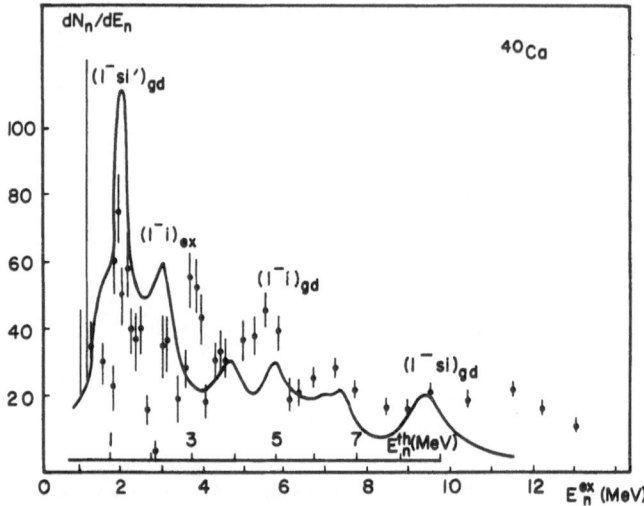

Fig. 15. Neutron spectrum following muon capture in ^{40}Ca, as calculated on the particle-hole model [48], compared to Schröder's [49] experimental data

Theoretical investigations on charged-particle decay channels of ^{12}B after muon capture in ^{12}C, such as

$$^{12}\text{B} \to {}^{11}\text{Be} + p\,, \tag{16a}$$

$$\to {}^{10}\text{B} + d\,, \tag{16b}$$

$$\to {}^{8}\text{Li} + \alpha\,, \tag{16c}$$

were made by Vartanyan et al. [51]. They showed that these processes also had to proceed via the giant resonance in order to explain large observed charged-particle emission rates [52].

The energy spectrum of charged particles following muon capture in ^{28}Si was measured by Sobottka and Wills [53]. The probabilities of the $(\mu^-, \nu p)$ reaction in ^{28}Si and ^{39}K were measured by Wilhelmova et al. [54] using an activation method with the results $(5.3 \pm 1.0)\%$ in ^{28}Si, $(3.2 \pm 0.6)\%$ in ^{39}K.

IV. Partial Decay Rates; Particle and Photon Emission

The daughter nuclei $^{A-A'}(Z-Z')$ resulting from the decay of the highly excited nucleus $^{A}(Z-1)$ created in muon capture, Eqs. (15b–e), correspond exactly to the type (n, p) and number of individual nucleons emitted in the decay $^{A}(Z-1)$, and they identify the decay channel about as well as the emitted nucleons do. Moreover, they are frequently left in an excited state. Much information may therefore be gained by observing their decay by photon emission,

$$^{A-A'}(Z-Z')^* \rightarrow {}^{A-A'}(Z-Z')^{(*)} + \gamma, \qquad (17)$$

especially in view of the fact that photon spectra are now very accurately measurable due to the availability of Ge(Li) detectors [55]. Muon capture experiments that observe subsequently emitted (delayed) nuclear gamma rays have therefore been undertaken recently with increased frequency. The following considerations render them most useful:

a) Each of the excited daughter nuclei in Eqs. (15) has a characteristic gamma decay cascade, Eq. (17).

b) The daughter nucleus $^{A-A'}(Z-Z')$, and hence often the decay channel in Eqs. (15) can be identified by this cascade.

c) Identification of the daughers and summing over the gammas provides us with knowledge of the distribution of final nuclei, and of the frequency of appearance of a given decay channel after muon capture (in particular, with the neutron multiplicities); this method, however, neglects the formation of the ground states of $^{A-A'}(Z-Z')$ which are not visible through a gamma decay.

d) The number of observed gammas compared to the number of stopped muons gives partial capture-decay rates for a given channel, and the sum over the channels in Eqs. (15b), (15c) etc. the muon capture rate to unbound states of $^A(Z-1)$.

e) Observation of decay-gammas in Eq. (15a) provides us with the muon capture probability without subsequent particle emission ("neutron-less muon capture"), i.e. the particle capture rates to all bound excited states of $^A(Z-1)$. [The rate for the ground state (plus the bound excited states) can however often be obtained by an observation of the beta decay back to ^AZ. For example, muon capture in ^{28}Si to all bound states ^{28}Al is determined by the subsequent beta decay of the ^{28}Al ground state back to the 1.78 MeV excited level of ^{28}Si, which can be observed by its gamma decay to the ^{28}Si ground state.]

f) Observation of the time decay rate of the gammas further identifies the process, and gives us the total muon capture rates.

Muon capture experiments with the observation of subsequently emitted gammas are due to Cramer and Telegdi [56] (time distribution only), Backenstoss [57], Acker and Backenstoss [58], Ehrlich [59], and Anderson [60]. Detailed and systematic studies have been carried out by Kaplan et al. [61] on the reaction

$$\mu^- + {}^{16}O \rightarrow \nu_\mu + {}^{16}N^*_{g.\,res.}$$
$$\phantom{\mu^- + {}^{16}O \rightarrow \nu_\mu + {}^{16}N^*_{g.}} \hookrightarrow {}^{15}N^* + n \qquad (18)$$
$$\phantom{\mu^- + {}^{16}O \rightarrow \nu_\mu + {}^{16}N^*_{g.res}} \hookrightarrow {}^{15}N + \gamma\,;$$

by Budyashov et al. [62] on muon capture in ^{12}C and ^{16}O; by Pratt [63] on ^{28}Si, ^{32}S, and ^{40}Ca; by Igo-Kemenes et al. [64] on ^{40}Ca; by Bunatyan et al. [65] on ^{27}Al, ^{28}Si, ^{51}V, and ^{56}Fe (neutron-less capture); by Backenstoss et al. [66] on Sc, Mn, Co, Nb, I, Bi, and Br; and by Earle and Bartholomew [67] on Zr, Ag, Au, Pb, and Bi. Furthermore, work is in progress by the groups at William and Mary (Miller, Welsh et al.) on ^{24}Mg and ^{28}Si (neutron-less) [68], at Berkeley (Kaplan, Temple et al.) on ^{27}Al, ^{28}Si and ^{59}Co, [69] and at Columbia (Cheng, Ruston et al.) on ^{209}Bi [70]. Finally, delayed gamma rays have also been observed after muon capture in ^{151}Eu and ^{153}Eu, by Petitjean [71].

Besides identifying the decay channels of the nucleus resulting from muon capture, Eqs. (15), by the gamma decay of their daughters, it is of course also possible to observe directly the emitted particles. This has been done for emitted neutrons, cf. Eqs. (15b, c, e), by MacDonald, Kaplan et al. [72], by using a cadmium-loaded liquid-scintillator tank. This method does not separate off the charged-particle decay channels, e.g. Eqs. (15d, e) etc.; these channels are small, however, see the end of Chapter III. The neutron-less decay channel, Eq. (15a), is not being directly observed here, either. This method is therefore complementary to the gamma-decay method in which all energetically allowed decay channels are seen, including the one without neutron emission, but in which all the ground-state decay channels are overlooked.

Some of the important experiments observing the decay of nuclei produced by muon capture will now be discussed in more detail.

a) Muon Capture without Neutron Emission, i.e. the reaction

$$\mu^- + {}^AZ \rightarrow {}^A(Z-1)^* + \nu_\mu$$
$$\phantom{\mu^- + {}^AZ \rightarrow} \hookrightarrow {}^A(Z-1)^{(*)} + \gamma\,. \qquad (19a)$$

This was studied by Budyashov et al. [62] and by Bunatyan et al. [65]. The former authors observed decay gammas in ^{12}B, from the process

$$\mu^- + {}^{12}C \rightarrow {}^{12}B^* + \nu_\mu$$
$$\phantom{\mu^- + {}^{12}C \rightarrow} \hookrightarrow {}^{12}B^{(*)} + \gamma\,. \qquad (19b)$$

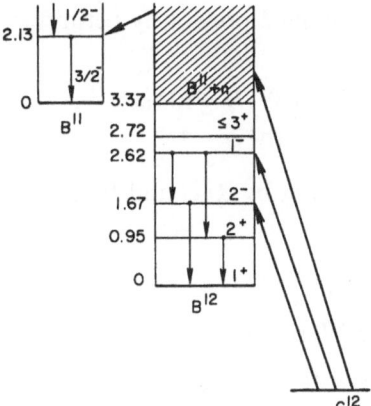

Fig. 16. Muon capture in ^{12}C leading to bound excited levels in ^{12}B, with subsequent gamma decay. Also shown is the gamma decay of ^{11}B produced by neutron emission [62]

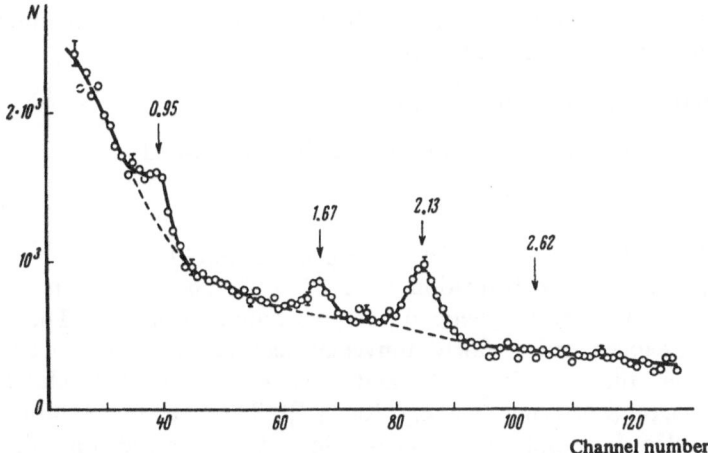

Fig. 17. Spectrum of photons observed after muon capture in ^{12}C [62]

The level scheme, with the observed decays, is shown in Fig. 16, together with the gamma decay of the 2.13 MeV level in ^{11}B (produced by neutron emission from the continuum of ^{12}B), which was also seen in this experiment. The gamma cascade spectrum is shown in Fig. 17 as observed in a NaI counter. Although here the $2^+ \rightarrow 1^+$ and $1^- \rightarrow 2^-$ decays accidentally coincide in energy, as do the $2^- \rightarrow 1^+$ and the $1^- \rightarrow 2^+$ decays, the known character of the cascade transitions obtained previously permitted to obtain the partial capture rates

$$\Lambda(1^-, 2.62\ \text{MeV}) = (0.72 \pm 0.17) \times 10^3\ \text{sec}^{-1},\tag{20a}$$

$$\Lambda(2^-, 1.67\ \text{MeV}) \leq 0.24 \times 10^3\ \text{sec}^{-1}.\tag{20b}$$

Theoretical rates are given by Ruel and Brennan [73] as

$$\Lambda(1^-) = 3 \times 10^3\ \text{sec}^{-1},\tag{20c}$$

$$\Lambda(2^-) = 1 \times 10^3\ \text{sec}^{-1},\tag{20d}$$

and by Maguire and Werntz [74] as

$$\Lambda(1^-) = (1.19 \pm 0.24) \times 10^3\ \text{sec}^{-1},\tag{20e}$$

$$\Lambda(2^-) = (0.51 \pm 0.10) \times 10^3\ \text{sec}^{-1}.\tag{20f}$$

The experimental rate to all bound excited states in ^{12}B is thus

$$\Lambda(\text{bound exc.}) = (0.76 \pm 0.14) \times 10^3\ \text{sec}^{-1},\tag{20g}$$

and since the capture rate to all bound states in ^{12}B (obtained by observing the beta decay of ^{12}B) is known as [75]

$$\Lambda(\text{bound}) = (7.01 \pm 0.27) \times 10^3\ \text{sec}^{-1},\tag{20h}$$

one has for the partial capture rate to the ^{12}B ground state:

$$\Lambda(\text{ground}) = (6.3 \pm 0.3) \times 10^3\ \text{sec}^{-1}.\tag{20i}$$

The excited-state transition rate is seen to be small ($\sim 12\%$).

The ground state transition is of $0^+ \rightarrow 1^+$ type, and is the prime example of a giant magnetic dipole resonance excitation. The giant $M1$ resonances are a fairly universal feature of nuclei, occurring throughout the periodic table, and have been seen most clearly in $180°$ electron scattering (see, e.g. Ref. [7]). They are not as universal as the electric dipole resonances, since they depend on spin-flip transitions in shell model states, which are absent for doubly-closed shell nuclei [76]. Due to their collectivity [77], they possess strong transition probabilities for muon capture as well as for photon and electro-nuclear reactions. The $M1$ states are often particle-stable (as in ^{12}C $-$ ^{12}B), and thus contribute to muon capture without neutron emission. They have been shown to play a prominent role in such reactions for the nuclei ^{28}Si $-$ ^{28}Al and ^{56}Mn, from the experiments of Bunatyan et al. [65]. Table 2 lists the results of these authors for muon capture without neutron emission in ^{27}Al, ^{28}Si, ^{51}V, and ^{56}Fe, identified by the characteristic activity of the daughter nuclei ^{27}Mg, ^{28}Al, ^{51}Ti, and ^{56}Mn in a NaI counter. It is seen that

Table 2. Rates of muon capture without neutron emission (in % of total capture rate)

Target	% exp [65]	% theory [18]	% exp [72]
^{27}Al	10 ± 1	6	8 ± 4
^{28}Si	28 ± 4	38	36 ± 6
^{51}V	10 ± 1	15	—
^{56}Fe	16 ± 3	24	19 ± 3

neutron-less muon capture in these nuclei is relatively large, especially in ^{28}Si and ^{56}Fe where the giant $M1$ transitions contribute. The theoretical results are due to Ref. [18], while the last column was obtained from an analysis of the results of Ref. [72], and should thus be corrected for the unobserved emission of charged particles; but other experiments (see below) indicate that the probability for charged particles is small, and negligible for heavier elements.

Coming back to the giant magnetic resonances, the mentioned theory of Bunatyan [18] showed that the principal contribution to the rates in ^{28}Si and ^{56}Fe is due to $0^+ \rightarrow 1^+$ transitions to 1^+ states in ^{28}Al and ^{56}Mn. The 1^+ level in ^{28}Al is the $T_3 = -1$ analog of a 1^+, $T_3 = 0$ level in ^{28}Si which was studied (together with similar $M1$ levels in ^{20}Ne, ^{24}Mg, ^{32}S, and ^{36}Ar) in 180° electron scattering performed at the Naval Research Laboratory by Fagg and collaborators [78]. Figure 18 shows the spectrum of $T = 1$, $J = 1^+$ levels in ^{28}Si seen in electron scattering and the corresponding analog states in ^{28}Al, and Fig. 19 presents the observed spectrum of scattered electrons. One usually has one or a few outstanding $M1$ resonances, except in ^{32}S (and ^{36}Ar) where the $M1$ strength is fragmented [79]. In contrast to the ^{12}C case where the big $M1$ state has its analog in the ^{12}B ground state, the analog $M1$ levels are excited states of ^{24}Na, ^{28}Al, ^{32}P, and ^{36}Cl and thus can be seen by their gamma decay. Their identification in some of these sd shell nuclei, and in other nuclei, has now been accomplished at Berkeley [69] and at William and Mary [68] in such a fashion. In Fig. 20, we show the Berkeley results on the gamma cascades from various daughter nuclei after muon capture in ^{28}Si, cf. Eqs. (15), (17), including the neutron-less transition to the $M1$ states in ^{28}Al. It is seen that the muon capture provides a fine instrument for a study of the giant magnetic resonances.

b) Study of the Giant Dipole SU_4 *Multiplet.* The giant dipole resonance forms an SU_4 multiplet [11] comprising the photonuclear giant dipole states as well as giant electric spin flip and magnetic quadrupole states (see, e.g. Ref. [7]), and the muon capture process proceeds

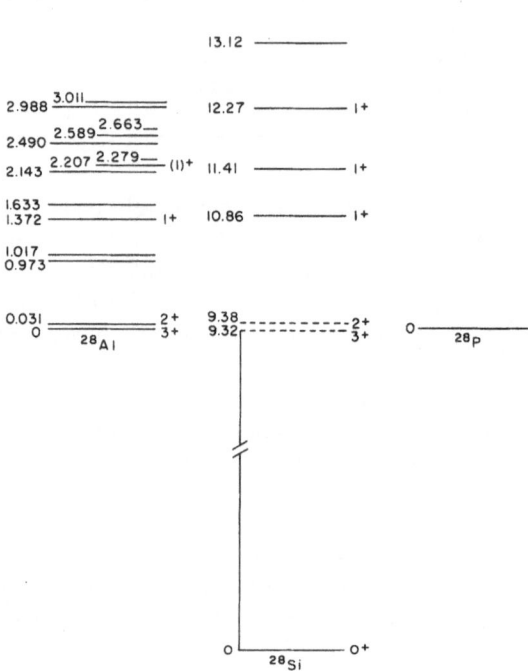

Fig. 18. Spectrum of *M* 1 (*T*= 1) levels in ²⁸Si, and analog states in ²⁸Al

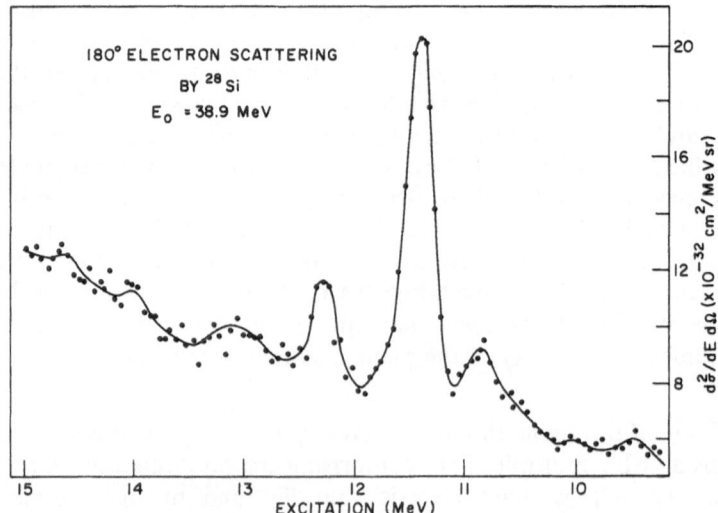

Fig. 19. Prominent M 1 states observed in the spectrum of 56 MeV electrons scattered at 180° from ²⁸Si [78]

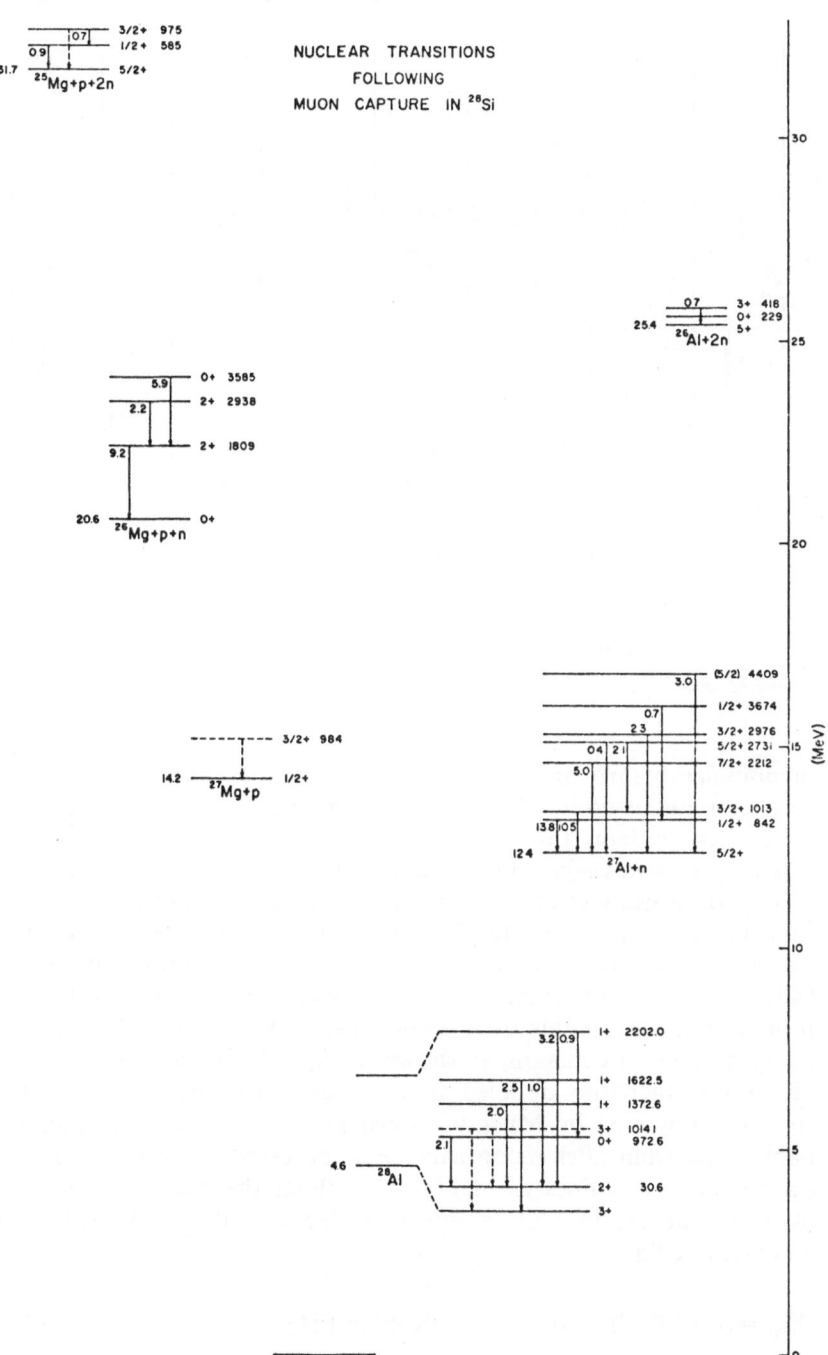

Fig. 20. Gamma cascades from various daughter nuclei following muon capture in
^{28}Si [69]

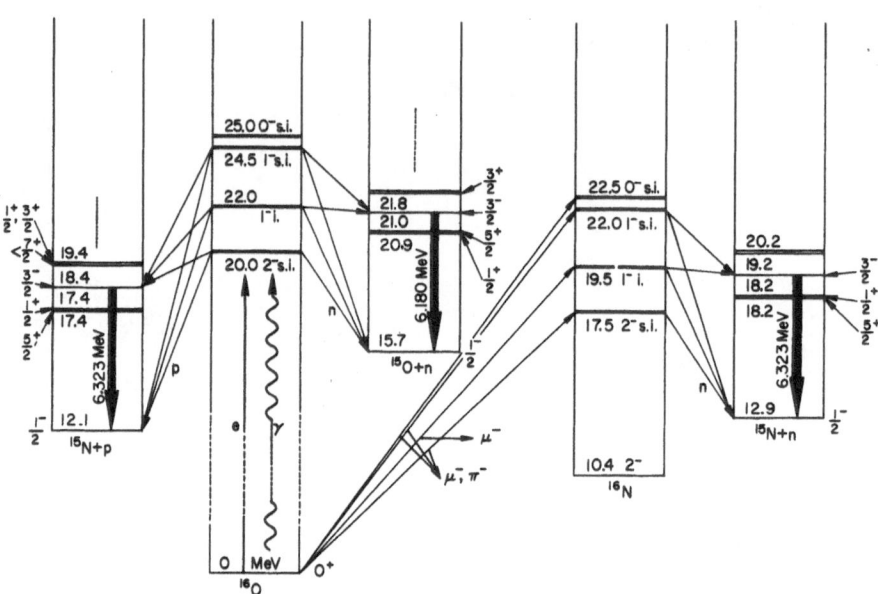

Fig. 21. Decay of giant resonances in ^{16}O excited by photons or electrons, and in ^{16}N excited by muons

predominantly through these states [11, 1]. Besides through the neutron spectra as discussed in Chapter III, this mechanism may be checked in an indirect fashion through the photon decay of the daughters in Eqs. (15). This was done in the reaction of Eq. (18) by Kaplan et al. [61] and by Budyashov et al. [62], by observing decay gamma rays in ^{15}N from the $\frac{3}{2}^-$, $\frac{1}{2}^+$, and $\frac{5}{2}^+$ levels to the ^{15}N ground state as shown in Fig. 6. Analogous decay photons were seen by Murray and Ritter [80] in the photoexcitation of ^{16}O with subsequent neutron or proton emission; in this case, analogous levels in both ^{15}N and ^{15}O decay by photon emission, as shown in Fig. 21. The number of decay photons is therefore doubled here, as demonstrated in the photon spectrum observed by Murray, shown in Fig. 22. The corresponding photon spectrum after muon capture as observed by Kaplan, Fig. 23, exhibits the same lines as in Fig. 22 without the doubling. The rate of muon capture in ^{16}O producing a decay of the $\frac{3}{2}^-$ level in ^{15}N was measured as

$$\Lambda_{exp.} = (2.5 \pm 0.23) \times 10^4 \sec^{-1} \quad \text{Kaplan [61]} \tag{21a}$$

$$= (2.0 \pm 0.5) \times 10^4 \sec^{-1} \quad \text{Budyashov [62]} \tag{21b}$$

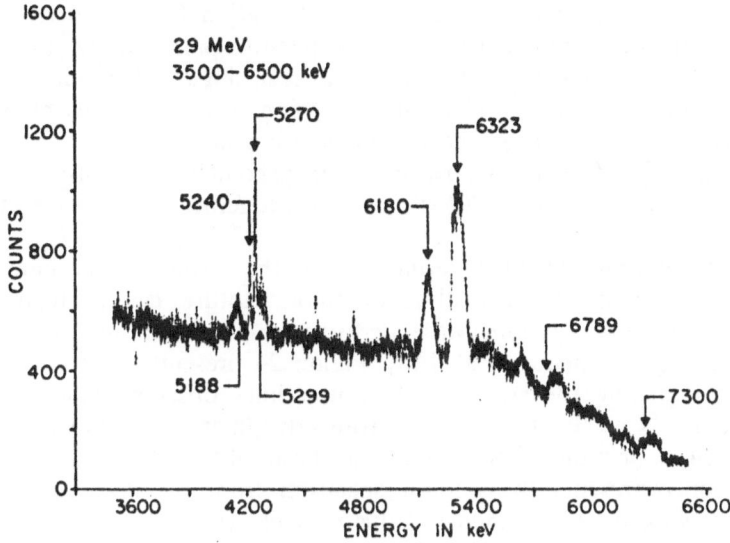

Fig. 22. Photon spectrum after photoexcitation of ^{16}O [80]

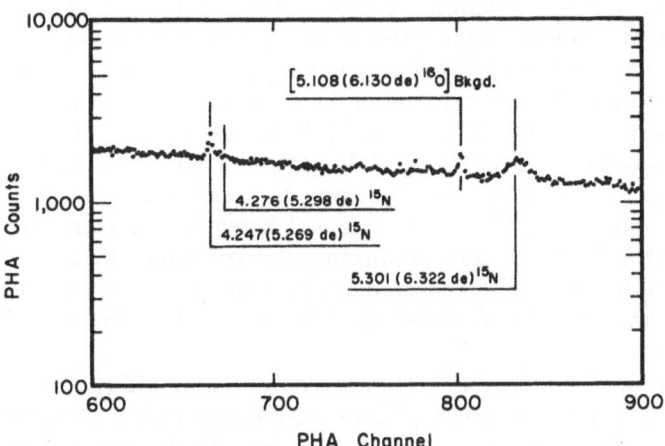

Fig. 23. Photon spectrum after muon capture in ^{16}O [61]

and may be compared to the theoretical rates obtained with the assumption of the giant resonance mechanism:

$$\Lambda_{theor.} = 5.2 \times 10^4 \, sec^{-1} \quad \text{Balashov [39]} \tag{21c}$$

$$= 3.0 \times 10^4 \, sec^{-1} \quad \text{Raphael [41]} \tag{21d}$$

The agreement provides an indirect confirmation for the correctness of the giant resonance mechanism. The positive parity decay channels in ^{15}N are not predicted in the single-particle, single hole model. Their experimental appearance gives an indication of the presence of two-particle, two-hole states in the ^{16}N giant resonance.

It might be of interest to perform an (n, γ) coincidence measurement, observing the predicted ~ 3 MeV neutrons together with the 6.3 MeV photon (Fig. 6).

A much more detailed verification of the resonance mechanism, using the gamma decay method of muon capture, can be made in ^{40}Ca where investigations were performed in this manner by Pratt [63] and Igo-Kemenes et al. [64]. Figure 24 presents Igo-Kemenes's results, depicting the strength of photon lines observed after muon capture in ^{40}Ca. The decay of the various daughter nuclei shown in the upper right portion of the figure has been observed, but the main figure only presents the photon spectrum from the decay of ^{39}K, together with the level scheme of this nucleus. The lines are compared to those obtained from ^{39}K in the (γ, γ') reaction in ^{40}Ca [81], shown in the right part of Fig. 24. The very detailed agreement of the lines which appear with the same relative intensities in both photonuclear and muon capture reactions, confirms the correctness of the giant resonance mechanism in the latter process, since it is known to hold in photonuclear reactions [82]. The figure depicts only the decay of states reached through the isospin part of the giant resonance, which in ^{40}Ca lies between 11.5 and 15.5 MeV [83]. Some of the spin flip (*si*) states lie much higher up, around 25 MeV [83], and are strongly excited in muon capture. Indeed, Igo-Kemenes has identified the copiously produced daughters ^{38}K, ^{38}Ar, ^{38}Cl which require an excitation of at least 22 MeV in ^{40}Ca, thus confirming the resonance mechanism via spin flip states. The observed gammas account for $\sim 40\%$ of all muon capture transitions, which shows the importance of the giant resonance mechanism.

It is seen, therefore, that a comparison of $(\mu, \nu\gamma)$ and (γ, γ') reactions verifies the giant resonance process and distinguishes between its *i* and *si* components. This comparison should be carried out by performing muon capture in more isotopes in which (γ, γ') results are now available, such as in ^{12}C [84–86], ^{14}N [87], ^{27}Al [88] or ^{32}S [87–89].

Such a comparison may also help decide the question of the isospin splitting of the giant resonance in $T \neq 0$ nuclei, possibly observed in (γ, n) reactions, e.g., in ^{26}Mg [90], or in (γ, γ') reactions, e.g., in ^{11}B [91, 92] or ^{13}C [93]. The giant resonance, e.g. in ^{11}B is expected theoretically [94, 95], to be split into a $T = T_< = \frac{1}{2}$ and a

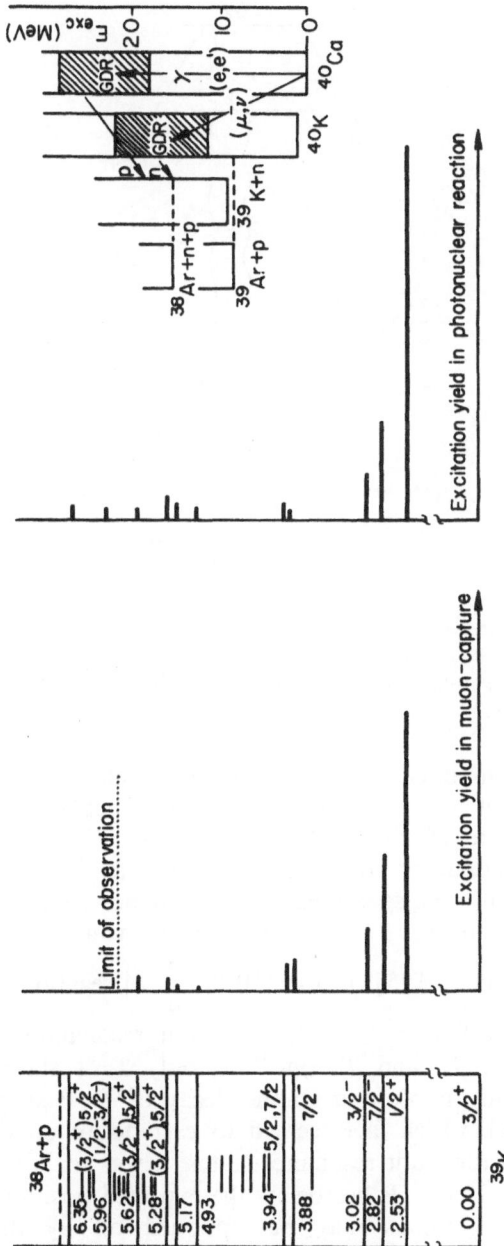

Fig. 24. Photon spectrum from the decay of ^{39}K produced by muon capture in ^{40}Ca [64], compared to the photon spectrum after photoexcitation of ^{40}Ca [81]

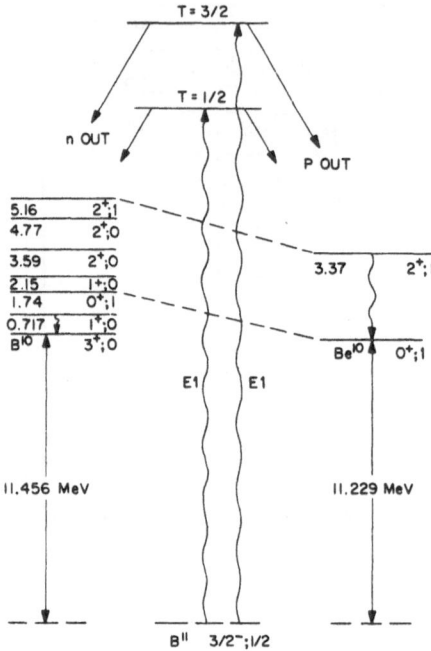

Fig. 25. Isospin splitting of the giant resonance in ^{11}B and $n(p)$ decay channels

$T = T_> = \frac{3}{2}$ component, which are separated cleanly, the $T_<$ resonance
lying at about 18 MeV and the $T_>$ resonance higher up, at about
22 MeV. Figure 25 shows this situation for $E\,1$ photoabsorption in
^{11}B, with subsequent $n(p)$ decay channels to ^{10}B (^{10}Be). In photo-
absorption, only the separated giant resonance peaks are visible [96];
an assignment of $T_<$, $T_>$ may be possible through a comparison with
the (p, γ_0) reaction [95]. In the ^{11}B $\left(\gamma, \dfrac{p}{n}\,\gamma'\right)$ reaction, the $T = 0$ states
in ^{10}B can be fed only by the $T_<$ giant resonance levels, the $T = 1$
states by both $T_>$ and $T_<$ (in ^{10}Be, only $T = 1$ states are possible).
Originally, only photons from the decay of $T = 0$ states in ^{10}B were
observed [91], which fact seemed to cast doubt on the existence of
isospin $T_>$ in the giant resonance states. It was realized, however, that
the pn threshold in ^{11}B lies below the expected position of $T_>$, so that
the expected gammas may be siphoned off by the pn channel.
A similar remark applies to the upper giant resonance in ^{26}Mg [90]
at ~ 22 MeV: its neutron decay to $T = \frac{1}{2}$ states in ^{25}Mg was not
observed, which seemed to indicate that it was $T = T_> = 2$; but again
the $2n$ threshold below the position of $T_>$ may siphon off a decay from

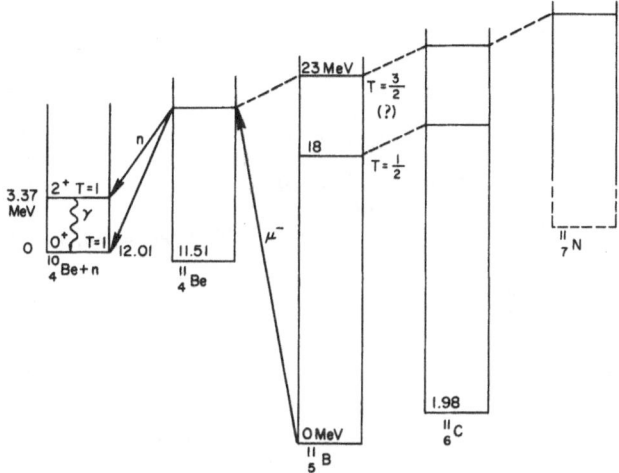

Fig. 26. Proposed muon capture experiment in ^{11}B to test isospin splitting in the giant resonance, with subsequent decay channels indicated

states that are really $T=1$ while being surmised to be the $T_>$ states. One sees that an identification of true $T_>$ states is not conclusive in these experiments; nor is this the case in Medicus' experiment in ^{11}B [92], where decay gammas from $T=1$ levels in ^{10}B were seen only for a bremsstrahlung endpoint of at least 25 MeV above the ^{11}B ground state, although this result makes the $T_>$ assignment to $E \gtrsim 25$ MeV states more likely.

We would like to propose here that muon capture experiments may shed more light on the question of isospin splitting in giant resonances. They may even give information on a possible isospin splitting of the spin flip components of the giant resonances, since these si states are strongly excited in muon capture while photonuclear experiments excite almost exclusively the i components of the resonances. We shall discuss the experiment for ^{11}B, although heavier $T \neq 0$ isotopes may be more favorable for muon capture on account of their larger capture rates.

Figure 26 depicts the i-type $T_< = \frac{1}{2}$ levels in ^{11}B and ^{11}C forming a doublet, and the surmised $T_> = \frac{3}{2}$ levels forming a quartet. The $T_3 = -\frac{3}{2}$ component of the latter in ^{11}Be represents the only i giant resonance channel (si channels can be considered in a similar fashion) for muon capture in ^{11}B, which should proceed strongly and decay to ^{10}Be with the emission of neutrons strongly peaked at 9 MeV, 5.6 MeV, etc., the latter transition emitting a 3.37 MeV gamma-ray in coincidence, which however may also be observed by itself. The

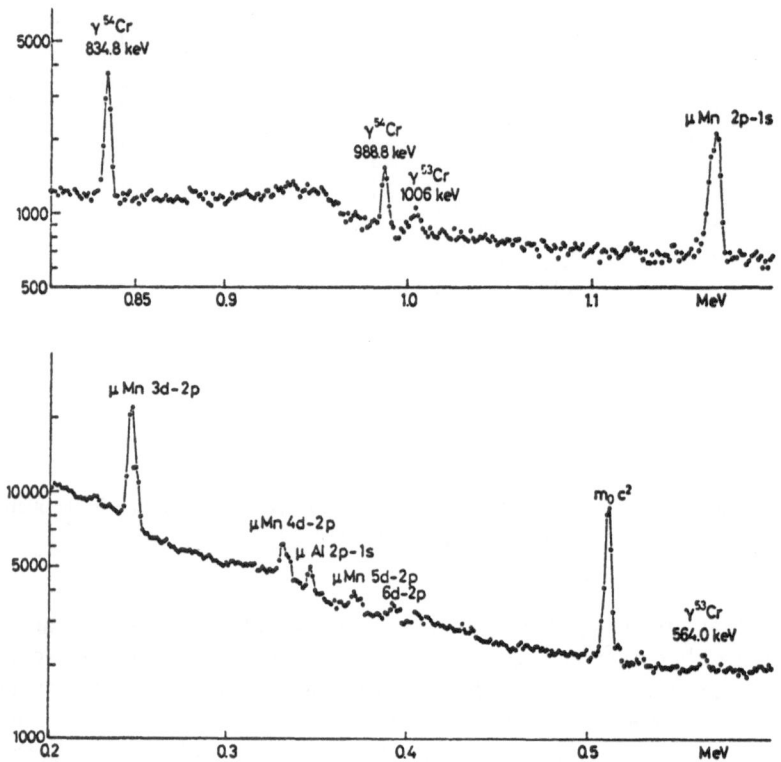

Fig. 27. Photon spectrum obtained from muon capture in ^{55}Mn [66]

partial rate of such a transition should constitute a sizable fraction of the total muon capture rate. If the upper giant resonance has $T=\frac{1}{2}$ (rather than $T_> = \frac{3}{2}$), it would have no analog in ^{11}Be and the mentioned neutron and gamma peaks would be absent, or at most contribute a small fraction of the total muon capture rate.

c) Neutron Multiplicities of neutrons emitted after muon capture have been measured both by observing the neutrons directly [72], as was done in capture by Al, Si, Ca, Fe, Ag, I, Au, and Pb, or by observing decay gammas of daughters [66], done in capture by ^{45}Sc, ^{55}Mn, ^{59}Co, ^{93}Nb, ^{127}I, and ^{209}Bi (keeping in mind that ground state transitions are not seen in this way). Backenstoss [66] identifies gamma cascades of many product nuclei. Parts of the gamma spectrum obtained from ^{55}Mn is shown in Fig. 27. Photon lines from cascades in ^{54}Cr, ^{53}Cr, and ^{52}Cr are identified (the line of ^{55}Cr is masked by an

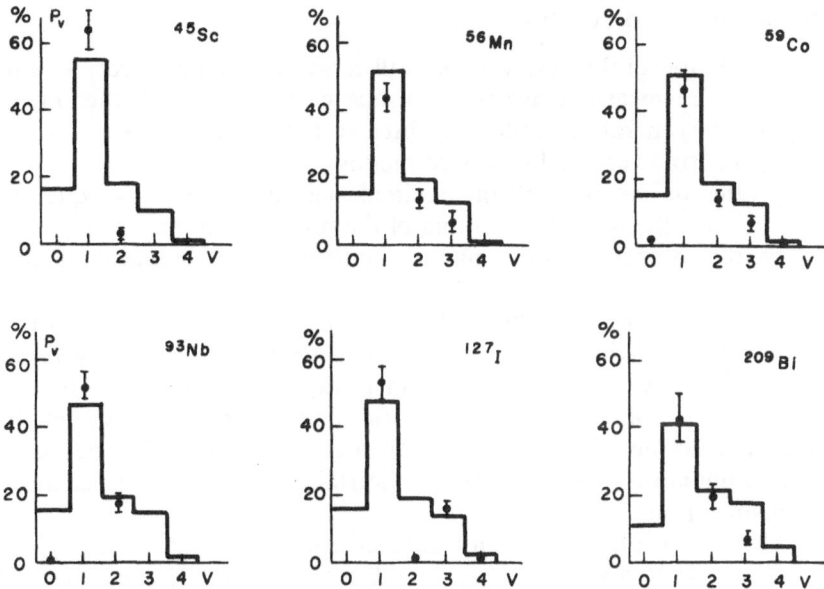

Fig. 28. Multiplicities of neutrons emitted after muon capture in various nuclei [66]. Solid lines are theoretical results of Singer [97]

atomic transition). It is interesting that no daughter nuclei corresponding to charged particle emissions were seen, in agreement with a general feature for heavier nuclei as mentioned earlier. Multiplicities of neutrons are obtained, as far as their emission leads to bound excited states of daughter nuclei. This includes the probabilities of no neutron emission (with the same reservation). Another point of caution is that certain decay channels may not be seen if their photon lines are masked by atomic gammas.

These results may be compared with the neutron multiplicities obtained by Kaplan [72] through a direct observation of emitted neutrons in a scintillator tank. (The latter method lumps into this the unobserved charged particle channels, but Ref. [66] indicated that these processes are unlikely). The results of Backenstoss lead to the neutron multiplicities shown in Fig. 28. One recognizes that the probabilities of zero-neutron emission are relatively small ($\lesssim 5\%$) while from Kaplan's [72] and Bunatyan's [65] results, somewhat larger probabilities for this decay channel had been indicated. Likewise, a 10–15% probability had been found theoretically by Singer [97]; the multiplicities of the latter work are shown as solid lines in Fig. 28.

V. Angular Correlations

The final topic of this review paper will concern the angular correlations
that may be measured between various particles emitted after muon
capture. In general, the following three vectors are available:

s_μ, the spin vector of polarized muons

p_ν, the momentum of the emitted neutrino (which is equal to
$-p_R$, the recoil momentum of the product nucleus), and

p_γ, the momentum of a photon emitted in a subsequent gamma
decay.

We do not consider here the momenta p_n of emitted neutrons;
correlations of the form $s_\mu \cdot p_n$ have already been dealt with in
Chapter II. It was mentioned after Eq. (21d), however, that a
measurement of p_n, p_γ coincidences (and correlations) would be
important to identify the energy of the neutron group that produced a
given gamma ray, and hence better ascertain the resonance mechanism
of muon capture.

General theories of (p_ν, p_γ) angular correlations have been given
by Popov [98], Oziewicz [99], and Bukhvostov [100]. The latest of
these theoretical papers include capture rate differences from the two
hyperfine states in a non-zero spin target, and a possible circular
polarization of photons (of degree η). The general capture-decay rate
including correlations is written as

$$\Lambda = \Lambda_0 \{ 1 + A s_\mu \cdot p_\nu + \eta B s_\mu \cdot p_\gamma + \cdots + E P_2(p_\nu \cdot p_\gamma) + F P_3(p_\nu \cdot p_\gamma) \} , \quad (22a)$$

with Λ_0 the uncorrelated rate, the coefficients being nuclear matrix
elements. For the special case of allowed transitions, unpolarized muon
and no observation of a circular polarization of photons, one has

$$\Lambda = \Lambda_0 \{ 1 + E P_2(p_\nu \cdot p_\gamma) \} . \quad (22b)$$

Bukhvostov [100] showed E to be a sensitive function of G_P/G_A,
while being insensitive to the nuclear matrix elements. For example,
E varies monotonically from 0 to -0.4 as G_P/G_A varies from 0 to 40
for muon capture from the lower hyperfine state in ^{10}B; but
$E \cong \text{const} = 0.1$ for capture from the upper hyperfine state. Measurement
of gamma-recoil angular correlations after muon capture is thus a
means for determining the induced pseudoscalar coupling constant G_P.

While it may be difficult to measure the direction of the recoiling
nucleus itself, an indirect method for its determination can be used if
the half-life of the excited state is less than 10^{-13} sec (or if a gas target
is used) so that the recoiling nucleus is not brought to rest too soon.
In this case, the line of the emitted gamma experiences a Doppler-
broadening due to the motion of the nucleus that emits the gamma.
The shape of the Doppler-broadened line depends sensitively on a

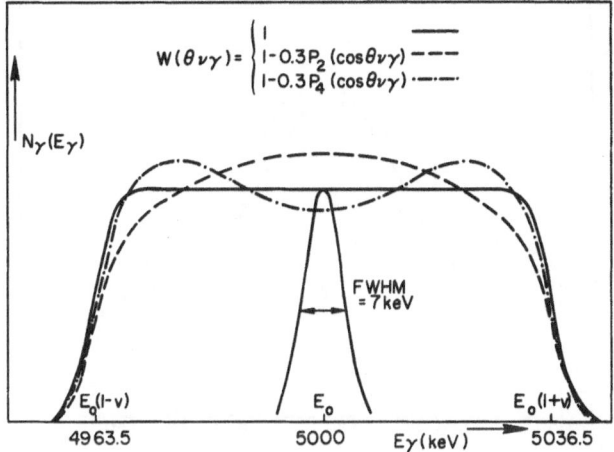

Fig. 29. Doppler-broadened line shapes for correlations $W = 1 - 0.3\,P_2$ and $1 - 0.3\,P_4$ [101]

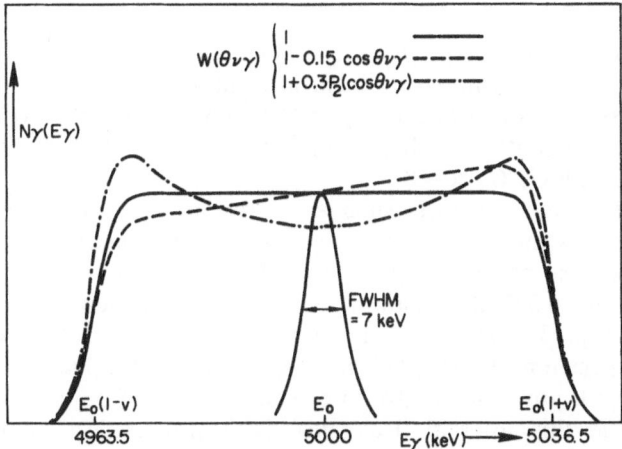

Fig. 30. Doppler-broadened line shapes for correlations $W = 1 - 0.15\,P_1$ and $1 + 0.3\,P_2$ [101]

correlation of the form of Eq. (22b), so that the correlation coefficient E may be determined from the line shape. This has been pointed out by Grenacs et al. [101], who plotted expected gamma line shapes for various assumed correlations. Figure 29 shows the broadened lines $W = \Lambda/\Lambda_0$ for a 5 MeV gamma ray and a recoil velocity $v = 0.0073c$ ($A \cong 14$), folding in a 7 keV resolution of the Ge (Li) detector. Assumed correlations of $1 - 0.3\,P_2(\boldsymbol{p}_v \cdot \boldsymbol{p}_\gamma)$ or $1 - 0.3\,P_4(\boldsymbol{p}_v \cdot \boldsymbol{p}_\gamma)$ are seen to lead to quite different line shapes. In Fig. 30, the correlations $1 - 0.15\,P_1$

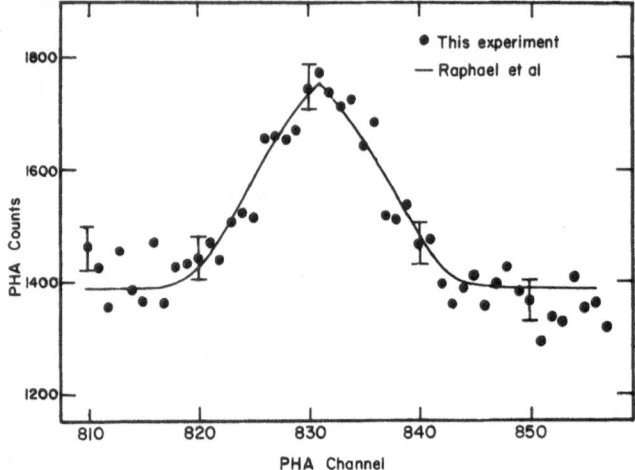

Fig. 31. Line shape fit of the Doppler-broadened 6.322 MeV gamma ray emitted after muon capture in ^{16}O [102]

$(\boldsymbol{p}_v \cdot \boldsymbol{p}_\gamma)$ and $1 + 0.3\, P_2(\boldsymbol{p}_v \cdot \boldsymbol{p}_\gamma)$ are shown. (With a P_1 term, the line shape becomes unsymmetric.) Such line shape measurements are now in progress at William and Mary [68] for the 1.231 MeV level in ^{28}Al. The observed butterfly shape suggests $E > 0$ in Eq. (22b) and is consistent with $G_P/G_A \cong +8$.

In a two-step emission, such as in reaction (18), line shapes may appear even if there is complete isotropy in the angular correlations. Using this assumption, Kaplan et al. [102] have fitted the observed shape of the 6.322 MeV gamma line in this reaction, with a result show in Fig. 31. This was achieved by folding in the theoretical neutron spectrum of Ref. [41]. It was found that the line shape could be fitted with any neutron group of energy between 3 and 6 MeV, but not outside this interval. Reference [41] predicted a predominant neutron group around 4 MeV (Fig. 10), which is also seen experimentally (Figs. 9, 12); the line shape of Fig. 31 is thus to a certain extent another confirmation of the resonance mechanism in muon capture.

The recoil effect is largest in a light nucleus. For muon capture in ^{4}He, in the reaction

$$\mu^- + {}^4\text{He} \rightarrow n + {}^3\text{H} + \nu_\mu, \tag{23}$$

one may measure the spectrum of the emitted ^{3}H, which in the laboratory is distorted depending on the angular correlation between the neutrino and the triton due to the large recoil of the intermediate ^{4}H. This was obtained theoretically by Crone and Werntz [103], assuming the resonance mechanism of muon capture and using a particle-hole model

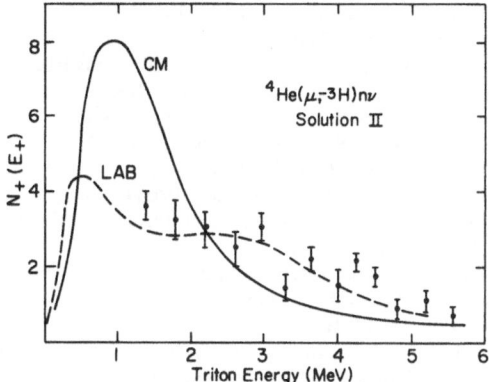

Fig. 32. Spectrum of tritons emitted after muon capture in ^{4}He in the CM system (solid line), and recoil-distorted laboratory spectrum (broken line) [103]. Fit of the latter spectrum to the data of Ref. [104]

for the ^{4}H giant resonances. The results for the ^{3}He spectrum are shown in Fig. 32, and the distortion of the spectrum due to the recoil and angular correlation effects is evident. (The solid curve is the CM spectrum, the dashed curve the laboratory spectrum.) The laboratory curve provides an excellent fit to the measured ^{3}He spectrum as obtained by Auerbach et al. [104].

The work of the author cited in this Review has been partially supported by the National Science Foundation.

References

1. Überall, H.: Proc. Williamsburg Conf. on Intermed. En. Phys., College of William and Mary, Williamsburg, Va., Feb. 10–12, p. 327, 1966. – See also Überall, H.: Suppl. Nuovo Ciment. **4**, 781 (1966).
2. Eramzhian, R. A.: Proc. Conf. on Electron Capture, Debrecen, Hungary, July 15–18, p. 269, 1968.
3. Überall, H.: Acta Phys. Austr. **30**, 89 (1969).
4. Überall, H.: In: Devons, S. (Ed.): High energy physics and nuclear structure, p. 48. New York: Plenum Press, 1970.
5. Überall, H.: Rep. No. 70–035, Center for Theor. Phys., Univ. of Maryland, Oct. 1970, and Proceed. NATO Adv. Inst. Electron Scatt. Nucl. Struc. (Santa Margherita di Pula, Cagliari, Italy, 1970), Bosco, B. (Ed.), Gordon and Breach, 1974 (to be published).
6. Bressani, T.: Excitation of isobaric analog states in meson-induced nuclear reactions. Rivista Nuovo Cimento. **1**, 268 (1971).
7. The muon and pion reactions mentioned here may be considered as important complements to photonuclear studies and to the inelastic electron scattering technique, which are exceedingly well-developed by now; see e.g. Ref. [1], or Überall, H.: Springer Tracts Mod. Phys. **49**, 1 (1969), and in particular Überall, H.: Electron scattering from complex nuclei, New York: Academic Press, 1971.

8. Kelly, F. J., McDonald, L. J., Überall, H.: Nucl. Phys. A **139**, 329 (1969). – Eisenberg, J. M., Weber, H. J.: Phys. Letters **34** B, 107 (1971).
9. Deutsch, J. P. et al.: Phys. Letters **26** B, 315 (1968). – Bistirlich, J. A., et al.: Phys. Rev. Letters **25**, 689, 960 (1970): Holland, M. M., Minehart, R. C., Sobottka, S. E.: Nucl. Phys. A **147**, 249 (1970). – Alder, J. C., et al.: Nuovo Cimento. Lettere **4**, 256 (1970); Hilscher, H., et al.: Nucl. Phys. A **158**, 584 (1970).
10. Primakoff, H.: Rev. Mod. Phys. **31**, 802 (1959).
11. Foldy, L. L., Walecka, J. D.: Nuovo Cimento **34**, 1026 (1964).
12. Luyten, J. R., Rood, H. P. C., Tolhoek, H. A.: Nucl. Phys. **41**, 236 (1963).
13. Sens, J. C.: Phys. Rev. **113**, 679 (1958).
14. Eckhause, M., Siegel, R. T., Welsh, R. E., Filippas, T. A.: Nucl. Phys. **81**, 575 (1966).
15. Landau, L. D.: Sov. Phys. JETP **8**, 70 (1959).
16. Migdal, A. B.: Sov. Phys. JETP **16**, 1366 (1963).
17. Migdal, A. B.: Theory of finite Fermi systems. New York: Interscience, 1967.
18. Bunatyan, G. G.: Sov. J. Nucl. Phys. **2**, 619 (1966); **3**, 613 (1966).
19. Novikov, V. M., Urin, M. G.: Sov. J. Nucl. Phys. **3**, 419 (1966).
20. Rho, M.: Phys. Rev. Letters **18**, 671 (1967).
21. Devanathan, V., Rho, M., Srinivasa Rao, K., Nair, S. C. K.: Nucl. Phys. B **2**, 329 (1967).
22. Gell-Mann, M.: Phys. Rev. **111**, 362 (1958).
23. Danos, M., Fuller, E. G.: Ann. Rev. Nucl. Sci. **15**, 29 (1965).
24. Wigner, E.: Phys. Rev. **51**, 106 (1937).
25. Goldhaber, M., Teller, E.: Phys. Rev. **74**, 1046 (1948).
26. Überall, H.: Phys. Rev. **137**, B 502 (1965).
27. Tomusiak, E. L.: Am. J. Phys. **36**, 1096 (1968).
28. Walker, G. E.: Phys. Rev. **151**, 745 (1966).
29. Brown, G. E., Castillejo, L., Evans, J. A.: Nucl. Phys. **22**, 1 (1961). – Lewis, F. H.: Phys. Rev. **134**, B 331 (1964).
30. Cannata, F., Leonardi, R., Rosa Clot, M.: Phys. Letters **32** B, 6 (1970).
31. Christensen, C. J., Krohn, V. E., Ringo, G. R.: Phys. Letters **28** B, 411 (1969).
32. Cannata, F.: Nuovo Cimento Lettere **4**, 75 (1970).
33. Yovanovich, N. L., Evseyev, V. S.: Phys. Lett. **6**, 332 (1963). – Evseyev, V. S., et al.: Sov. J. Nucl. Phys. **4**, 245, 378 (1967).
34. Sundelin, R. M., et al.: Phys. Rev. Letters **20**, 1198, 1201 (1968); Sculli, J., et al.: Bull. Am. Phys. Soc., **13**, 678 (1968).
35. Piketty, C. A., Procureur, J.: Nucl. Phys. B **26**, 390 (1971). – Eramzhyan, R. A., Fetisov, V. N., Salganic, Yu. A.: Phys. Rev. Letters **35** B, 143 (1971).
36. Dolinskii, E. I., Blokhintsev, L. D.: Sov. Phys. JETP **8**, 104 (1959); **12**, 1260 (1961); Devanathan, V., Rose, M. E.: J. Math. Phys. Sci. (India) **1**, 137 (1967).
37. Klein, R., Neal, T., Wolfenstein, L.: Phys. Rev. **138**, B 86 (1965).
38. Bogan, A.: Phys. Rev. Letters **22**, 71 (1969); Phys. Rev. B **12**, 89 (1969).
39. Balashov, V. V., Beliaev, V. B., Eramjian, R. A., Kabachnik, N. M.: Phys. Letters **9**, 168 (1964).
40. Donnelly, T. W.: Nucl. Phys. A **150**, 393 (1970). – Balashov, V. V., Kabachnik, N. M., Markov, V. I.: Nucl. Phys. A **129**, 369 (1969).
41. Raphael, R., Überall, H., Werntz, C.: Phys. Letters **24** B, 15 (1967).
42. Kelly, F. J., Überall, H.: Nucl. Phys. A **118**, 302 (1968).
43. Lewis, F. H., Walecka, J. D.: Phys. Rev. **133**, B 849 (1964). – de Forest, T.: Phys. Rev. **139**, B 1217 (1965).
44. Kamimura, M., Ikeda, K., Arima, A.: Nucl. Phys. A **95**, 129 (1967).
45. Eveseyev, V., et al.: Phys. Letters **28** B, 553 (1969).
46. Evseyev, V., Kozlowski, T., Roganov, V., Woitkowski, J.: In: Devons, S. (Ed.): High energy physics and nuclear structure, p. 157. New York: Plenum Press, 1970.

47. Plett, M. E., Sobottka, S. E.: Phys. Rev. C3, 1003 (1971).
48. Hill, L. L., Überall, H.: Nucl. Phys. A190, 341 (1972).
49. Schröder, W. U.: Thesis, Techn. Hochschule Darmstadt (1971). – Jahnke, V. et al.: Submitted to the 1971 High Energy and Nuclear Structure Conference, Dubna.
50. Überall, H.: Phys. Rev. 139, B1239 (1965).
51. Vartanyan, V. A., Zhusupov, M. A., Eramjian, R. A.: Dubna preprint P4–4314 (1969).
52. Morinaga, H., Fry, W. F.: Nuovo Cimento 10, 308 (1953). – Vaisenberg, A. O. et al.: Yad. Fiz., 1, 652 (1965). – Komarov, V. I. et al.: Dubna preprint P 1–3721 (1968). – Kotelchuk, D., Tyler, Y. V.: Phys. Rev. 165, 1190 (1968).
53. Sobottka, S. E., Wills, E. L.: Phys. Rev. Letters 20, 596 (1968).
54. Wilhelmova, L., et al.: Sov. J. Nucl. Phys. 13, 310 (1971).
55. Ewan, G. T., Tavendale, A. J.: Can. J. Phys. 42, 2286 (1964).
56. Cramer, W. A., Telegdi, V. L., Winston, R., Lundy, R. A.: Nuovo Cimento 24, 546 (1962).
57. Backenstoss, G., et al.: Nucl. Phys. 62, 449 (1965).
58. Acker, H., Backenstoss, G., Daum, C., Sens, J., de Wit, S.: Nucl. Phys. 87, 1 (1966).
59. Ehrlich, R. D. et al.: Phys. Rev. Letters 18, 959 (1967); 19, 344 (1967).
60. Anderson, H. L. et al.: Phys. Rev. 187, 1565 (1969).
61. Kaplan, S. N., Pyle, R. V., Temple, L. E., Valby, G. F.: Phys. Rev. Letters 22, 795 (1969).
62. Budyashov, Yu. G. et al.: Sov. Phys. JETP 31, 651 (1970).
63. Pratt, T. A. E.: Nuovo Cimento 61B, 119 (1969).
64. Igo-Kemenes, P., Deutsch, J. P., Favart, D., Grenacs, L., Lipnik, P., Macq, D. C.: Phys. Letters 34B, 286 (1971).
65. Bunatyan, G. et al.: Sov. J. Nucl. Phys. 9, 457 (1969); 11, 444 (1970).
66. Backenstoss, G. et al.: Nucl. Phys. A162, 541 (1971). – Povel, H. P. et al.: Phys. Letters 33B, 620 (1970).
67. Earle, E. D., Bartholomew, G. A.: Nucl. Phys. A176, 363 (1971).
68. Miller, G. H., Eckhause, M., Martin, P., Welsh, R. E.: Phys. Rev. C6, 487 (1972).
69. Temple, L. E., Kaplan, S. N., Pyle, R. V., Valby, G. F.: Lawrence Berkeley Laboratory preprint LBL–24 (1971).
70. Rushton, A. M. et al.: Bull. Am. Phys. Soc. 16, 652 (1971).
71. Petitjean, C. et al.: Nucl. Phys. A178, 193 (1971).
72. MacDonald, B., Diaz, J. A., Kaplan, S. N., Pyle, R. V.: Phys. Rev. 139, B1253 (1965).
73. Ruel, M., Brennan, J. G.: Phys. Rev. 129, 866 (1969).
74. Maguire, W., Werntz, C.: Nucl. Phys. A205, 211 (1973).
75. Maier, E. J., Edelstein, R. M., Siegel, R. J.: Phys. Rev. 133, 663 (1964).
76. Kurath, D.: Phys. Rev. 130, 1525 (1962).
77. Überall, H.: Bull. Am. Phys. Soc. 16, 561 (1971).
78. Fagg, L. W., Bendel, W. L., Jones, E. C., Numrich, S.: Phys. Rev. 187, 1378 (1969).
79. Fagg, L. W., Bendel, W. L., Cohen, L., Jones, E. C., Kaiser, H. F., Überall, H.: Phys. Rev. C4, 2089 (1971).
80. Murray, K. M., Ritter, J. C.: Phys. Rev. 182, 1097 (1969).
81. Ullrich, H., Krauth, H.: Nucl. Phys. A123, 641 (1969).
82. see, e.g. Danos, M., Fuller, E. G.: Ann. Rev. Nucl. Sci. 15, 29 (1965).
83. Hill, L. L.: Phys. Letters 25B, 169 (1967).
84. Maison, J. M.: Thesis, Univ. of Paris (Orsay) 1969.
85. Stewart, R. J. J., Thompson, M. N., Thomson, J. E. M.: Univ. of Melbourne, preprint UM-P-70/22 (1970).
86. Medicus, H. A., Bowey, E. M., Gayther, D. B., Patrick, B. H., Winhold, E. J.: Nucl. Phys. A156, 257 (1970).
87. Thompson, M. N., Steward, R. J. J., Thomson, J. E. M.: Phys. Letters 31B, 211 (1970).
88. Thomson, J. E. M.: Progress Report. School of Physics, Univ. of Melbourne, 1970.
89. Ishkhanov, B. S. et al.: Sov. J. Nucl. Phys. 12, 121 (1971).

90. Wu, C. P., Firk, F. W. K., Berman, B. L.: Phys. Letters **32** B, 675 (1970).
91. Murray, K. M.: Phys. Rev. Letters **23**, 1461 (1969). − Hayward, E., Schwartz, R. B., Murray, K. M.: Phys. Rev. C **2**, 761 (1970).
92. Patrick, B. H., Medicus, H. A., Mehta, G. K., Bowey, E. M., Gayther, D. B.: Phys. Letters **34** B, 488 (1971).
93. Murray, K. M., Toms, M. E.: Nuovo Cim. Lettere **1**, 571 (1971).
94. Fraser, R. F.: Thesis. Univ. of Melbourne (1968); also quoted in Ref. [95].
95. Kuan, H. M., Hasinoff, M., O'Connell, W. J., Hanna, S. S.: Nucl. Phys. A **151**, 129 (1970).
96. Hayward, E., Stovall, T.: Nucl. Phys. **69**, 241 (1965).
97. Singer, P.: Nuovo Cimento **23**, 669 (1962).
98. Popov, N. P.: Sov. Phys. JETP **17**, 1130 (1963). − Bukat, G. M., Popov, N. P.: Sov. Phys. JETP **19**, 1200 (1964).
99. Oziewicz, Z., Pikulsi, A.: Acta Phys. Polon. **31**, 501 (1967); Oziewicz, Z., Popov, A.: Phys. Letters **15**, 273 (1965).
100. Bukhvostov, A. P., Popov, N. P.: Phys. Letters **24** B, 497 (1967); Sov. J. Nucl. Phys. **6**, 589, 903 (1968).
101. Grenacs, L., Deutsch, J. P., Lipnik, P., Macq, P. C.: Nucl. Instrum. Meth. **58**, 164 (1968).
102. Kaplan, S. N., Pyle, R. V., Temple, L. E., Valby, G. F.: In: Devons, S. (Ed.): High energy physics and nuclear structure, p. 163. New York: Plenum Press, 1970.
103. Crone, L., Werntz, C.: Nucl. Phys. A **134**, 161 (1969).
104. Auerbach, L. B. et al.: Phys. Rev. **138**, B 127 (1965).

Prof. Dr. H. Überall
Physics Department, Catholic University, Washington, D. C.
and
U. S. Naval Research Laboratory, Washington, D. C.

Emission of Particles Following Muon Capture in Intermediate and Heavy Nuclei*

PAUL SINGER

Contents

I. Introduction

The purpose of this talk is to present a review of theoretical work and experimental results of recent years on the emission of particles from nuclei, following muon capture. As emphasized in the title, my lecture is restricted to intermediate and heavy nuclei, so as to minimize the overlap with Professor Überall's report [81]. Even so, for the sake of continuity, I am afraid that both of us will have to "intrude" occasionally into the other's domain.

The topic of my review is, I believe, customarily classified today as belonging to "intermediate energy nuclear physics" and has been succinctly covered in conferences dealing with "High Energy Physics and Nuclear Structure". Indeed, the study of the particles emitted after muon capture provides an additional angle for obtaining information on nuclear structure. The usefulness of the muon in this respect is enhanced when considered together with the analysis of other "particle physics tools", like K^- and π^- meson capture, pion photoproduction from nuclei, etc.

* Review talk presented at the Muon Physics Conference, Colorado State University, Fort Collins, Colorado, 6–10 September, 1971 (Expanded and updated version, September, 1973).

II. Neutron Emission

2.1 General

The basic process we consider is the capture of a muon by a nucleus from a K-atomic orbit. As a result of the weak interaction the following nuclear reaction occurs:

$$\mu^- + A_Z^N \rightarrow \nu_\mu + X . \tag{1}$$

The detectable product X consists of a residual heavy nucleus and light particles. In intermediate and heavy nuclei, the light particles are neutrons and (or) γ's in most of the cases. The few percent of charged light particles observed are mainly protons, but deuterons and α-particles have also been observed in still smaller quantities. In the following, we address ourselves to the question of neutron emission.

The average number of neutrons emitted per capture increases with the atomic number. The experimental figures [1] for a sample of naturally occurring elements ranging from Al to Pb are as follows:

Table 1

	Al	Si	Ca	Fe
Average number of neutrons per capture	1.26 ± 0.06	0.86 ± 0.07	0.75 ± 0.03	1.12 ± 0.04
	Ag	I	Au	Pb
	1.61 ± 0.06	1.44 ± 0.06	1.66 ± 0.04	1.71 ± 0.07

The increase in the average multiplicity of neutrons is however only a rough description, and the deviations from a smooth line, due to particular nuclear structure effects, are quite large.

Bobodyanov has pointed out [83] that the general trend of ν (= average number of neutrons per capture) is described well by the empirical function $\nu = (0.3 \pm 0.02) \, A^{1/3}$.

The emission of neutrons can be approximately classified as direct or from an intermediate "compound nucleus" formed after the muon capture process. Direct emission refers to the neutron created in the elementary process

$$\mu^- + p \rightarrow n + \nu_\mu , \tag{2}$$

which succeeds in leaking out of the nucleus. These neutrons have fairly high energies, from a few MeV to as high as 40–50 MeV [2–5].

The direct neutrons are expected to carry with them information on the basic process (2), like angular asymmetry with respect to the muon spin as a result of parity violation in the weak interaction.

Most of the neutrons emitted after capture seem however to be "evaporation neutrons". In intermediate and heavy nuclei the excitation energy acquired by the neutron formed in the capture process is shared with the other nucleons of the nucleus and a "compound nucleus" is formed. This intermediate excited nuclear state then loses energy by boiling-off mainly low-energy neutrons.

Some authors (e.g. Ref. [3]) distinguish between direct neutrons, evaporation neutrons and giant-resonance neutrons. The latter are the neutrons emitted following a capture process in which transitions to giant multipole nuclear levels play an important role [6, 7]. This manifestation of the nuclear giant resonance levels and the relation of muon capture to photoproduction by using the conserved vector current theory has been considered mainly with reference to light and intermediate nuclei [8]. The possibility of detecting typical "giant-resonance" neutrons in particular in intermediate and heavy nuclei, is however not a necessary corollary of the fact that the capture mechanism proceeds through these collective states. The de-excitation of the giant-multipole states, which occur in intermediate nuclei at excitation energies of 15–25 MeV, will most probably occur through a statistical boiling-off process. In Ca^{40} for instance, where the giant-dipole states are responsible for approximately 55% of the capture rate [7], some 80–90% of the neutrons emitted follow an evaporation spectrum, on top of which there is also evidence [5, 9] of a line spectrum indicative of transitions between resonant ^{40}K to excited ^{39}K. In lighter nuclei, it would be more feasible to detect line transitions from the giant resonance states.

2.2 Neutron Evaporation and Neutron Multiplicities

If the process (2) occurs with a proton at rest, the resulting neutron carries an energy of $E_n \simeq \mu^2 c^2/2M_n \simeq 6$ MeV. The nuclear protons have however a finite momentum distribution, and the neutrons can therefore emerge from the capture reaction with a spectrum of energies. In order to account for the low energy neutrons emitted after muon capture, the following physical picture [10] involving a two-step process can be used: the muon is captured by a quasi-free nucleon, whose acquired energy is distributed among the nucleons of the nucleus, and a compound nucleus is thus formed:

$$\mu^- + A_Z^N \rightarrow \nu_\mu + (A_{Z-1}^{N+1})^*$$ (3)

The excited nuclear state then loses energy by evaporating nuclear particles (mainly neutrons) and γ-rays till a ground state is reached.

$$(A_{Z-1}^{N+1})^* \rightarrow A_{Z-1}^{N+1} + \gamma\text{'s} \tag{4a}$$

$$\rightarrow A_{Z-1}^{N} + n + \gamma\text{'s} \tag{4b}$$

$$\rightarrow A_{Z-1}^{N-1} + 2n + \gamma\text{'s} \quad \text{etc.} \tag{4c}$$

This picture is expected to be of greater validity for heavier nuclei.

McDonald, Kaplan, Diaz, and Pyle [1, 11] have done the first extensive experiments to measure the neutron multiplicity distributions from μ^- capture in several intermediate and heavy nuclei. In their experiment, the neutrons emitted after μ^- capture are first thermalized and then detected by using a cadmium-loaded liquid-scintillator tank. In Fig. 1 we give the results of McDonald, Kaplan, Diaz, and Pyle for the emission of 0, 1, 2, ... neutrons after muon capture in various nuclei. The F_i are the multiplicities corrected for the experimental detection efficiency of the neutron counter (54.5%).

If the degenerate Fermi gas picture is used for the nucleons, the resultant excitation gives an average emission of neutrons of about half the observed one [11]. Moreover, no evaporation of 3 or 4 neutrons, as observed to occur in a certain percentage, is then accountable. A more realistic description of the capturing nucleus is obviously needed in order to calculate the nuclear excitation. As the process under consideration is mainly a volume effect, the Brueckner picture [12] for the constant-density region of the heavy nuclei can be used [13]. The nucleons are described as moving most of the time independently in a momentum-dependent average potential $V(p)$ created by the other nucleons. The momentum dependence of $V(p)$ is nearly quadratic for momenta lower than the Fermi momentum. The capturing nucleon can then be treated by using an effective mass

$$M^*(p) = p/(dE/dp).$$

From the $E(p)$ calculated for nuclear matter [12], one deduces for $p < p_F$

$$M^* = M\,(0.60 + 0.13\,p^2/p_F^2),$$

while for $p > p_F$, $M^*(p) \rightarrow M$. For finite nuclei, the effective mass might be even smaller [14]. The use of the effective mass approximation thus accounts for an increased excitation energy from the capture process.

A second improvement to be made is the use of a more realistic momentum distribution for the nucleons in the nucleus. Evidence from a wide range of phenomena reveals [13] that the nucleon

Target	Average multiplicity	Multiplicity distribution (adjusted to 0.545 efficiency)							
		F_0	F_1	F_2	F_3	F_4	F_5	F_6	F_7
Al	1.262 ± 0.059	0.449 ± 0.027	0.464 ± 0.028	0.052 ± 0.013	0.036 ± 0.007	−0.0023 ± 0.004	−0.001 ± 0.004	0.003 ± 0.004	
Si	0.864 ± 0.072	0.611 ± 0.042	0.338 ± 0.042	0.045 ± 0.018	−0.002 ± 0.008	0.003 ± 0.005	0.002 ± 0.005	0.003 ± 0.006	
Ca	0.746 ± 0.032	0.633 ± 0.021	0.335 ± 0.022	0.025 ± 0.009	0.004 ± 0.006	0.003 ± 0.003			
Fe	1.125 ± 0.041	0.495 ± 0.018	0.416 ± 0.019	0.074 ± 0.011	0.014 ± 0.005	−0.0001 ± 0.003	0.002 ± 0.003		
Ag	1.615 ± 0.060	0.360 ± 0.021	0.456 ± 0.023	0.144 ± 0.017	0.031 ± 0.009	0.007 ± 0.005	0.002 ± 0.004	0.001 ± 0.003	
I	1.436 ± 0.056	0.396 ± 0.021	0.474 ± 0.023	0.087 ± 0.015	0.035 ± 0.009	0.007 ± 0.005	0.0002 ± 0.004		
Au	1.662 ± 0.044	0.370 ± 0.015	0.425 ± 0.016	0.156 ± 0.012	0.032 ± 0.006	0.014 ± 0.004	0.003 ± 0.003	0.0003 ± 0.003	
Pb	1.709 ± 0.066	0.324 ± 0.022	0.483 ± 0.025	0.137 ± 0.018	0.045 ± 0.010	0.011 ± 0.006			
Ag	1.60 ± 0.18	0.389 ± 0.100	0.455 ± 0.075	0.120 ± 0.035	0.030 ± 0.015	0.001 ± 0.003	0.009 ± 0.006	0.000 ± 0.007	0.010 ± 0.007
Pb	1.64 ± 0.16	0.348 ± 0.100	0.479 ± 0.057	0.137 ± 0.027	0.018 ± 0.012	0.010 ± 0.005	0.005 ± 0.004	0.003 ± 0.004	0.002 ± 0.002

Fig. 1. The multiplicity distribution and the average neutron emission from the experiment of MacDonald et al. [1]. The last two items of the table refer to their previous experiment [11].

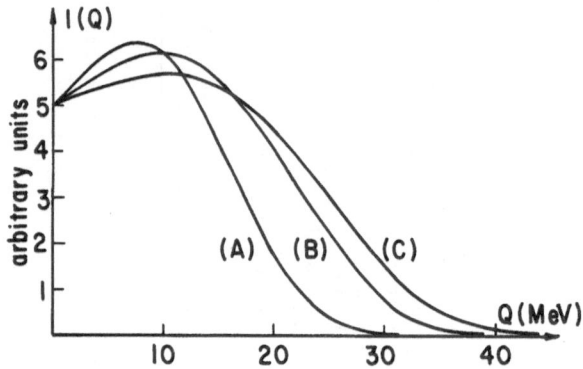

Fig. 2. The nuclear excitation distribution in Ag^{107}, calculated [13] for three choices of the effective mass and the Gaussian momentum distribution parameter: (A) $M^* = M$; $\alpha^2/2M = 14$ MeV. (B) $M^* = 0.68\,M$; $\alpha^2/2M = 14$ MeV. (C) $M^* = 0.68\,M$; $\alpha^2/2M = 20$ MeV

momentum distribution $\varrho(p)$ is well approximated by

$$\varrho(p) = N \exp\left[-p^2/\alpha^2\right], \tag{5}$$

with

$$15\text{ MeV} < \alpha^2/2M < 20\text{ MeV}.$$

The above two effects were taken [13] into account in calculating the nuclear excitation distribution $I(Q)$, and the result for Ag^{107} is shown in Fig. 2. In the calculation, the excitation energy

$$Q = {}_A(Mc^2)^Z - {}_A(Mc^2)^{Z-1} + \mu c^2 - B_\mu - k_\nu c \tag{6}$$

is related to the capturing process by

$$Q = (2M^*)^{-1}(q^2 - p^2), \tag{7}$$

where q, p are the momenta of the neutron and proton involved in (1). The Pauli principle effect is taken into account in the calculation of $I(Q)$. In Fig. 3 the difference between the excitation functions in Au^{197} obtained [1] with various momentum distributions for the nucleons is exhibited (the effective masses used were chosen so as to give the observed *average* neutron emission).

The emission of neutrons from the excited nucleus is calculated [1, 11, 13] by using the statistical theory of Weisskopf and Ewing [15]. The missing accurate knowledge of the level density of the nuclei involved has been replaced in the above calculations by the assumption of an evaporation spectrum for the emitted neutrons

$$dN(E) \sim E \exp\left[-E/\theta\right] dE. \tag{8}$$

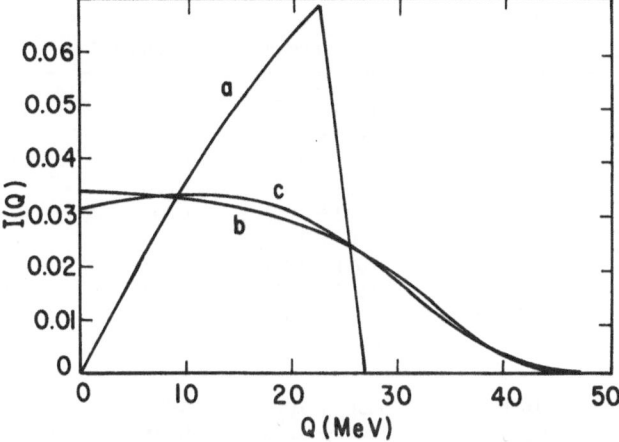

Fig. 3. The nuclear excitation distribution in Au, calculated [1] for various effective masses and momentum distributions: a $M^* = 0.38\,M$; Fermi gas, $\theta_f = 0$. b $M^* = 0.25\,M$; Fermi gas, $\theta_f = 12$ MeV. c $M^* = 0.46\,M$; Gaussian, $\alpha^2/2M = 20$ MeV

Using information from nuclear reactions at comparable energies, constant nuclear temperatures were used in the calculations [1, 13] with $\theta = 0.65 - 0.75$ MeV. Small changes in θ do not significantly affect the calculations.

Using a Gaussian momentum distribution with $\alpha^2/2M = 20$ MeV and an average effective mass of $M^* = 0.68\,M$, one finds that the theoretical prediction [13] is only in rough agreement with experiment (Fig. 4). The calculated average emission is generally 20% lower than observed and the agreement with experiment for f_0 and f_1, is generally poor.

One can improve the agreement by considering smaller effective nucleon masses. If the effective mass for Ag, I, Au, Pb is chosen to be $M^* \simeq 0.5\,M$, then the observed average neutron emission is well reproduced [1], but the multiplicity distributions calculated are still not in good agreement with experiment, especially for f_0 and f_1 (Fig. 5).

In the work of Kaplan et al. [1, 11] there is no experimental distinction between direct and evaporated neutrons. Hence, in comparing with their experiment, one should include in the calculation the appropriate percentage of neutrons directly emitted. When this is done, by assuming approximately 15–20% probability for neutrons to be emitted directly [13], the agreement of the calculation with experiment is generally improved. In order to reproduce the observed average neutron emission an effective mass $M^* \simeq 0.4\,M$ is now required in

Neutron multiplicities	Ag		I	
	Calculated	Experimental	Calculated	Experimental
f_0	0.488	0.383 ± 0.025	0.492	0.393 ± 0.026
f_1	0.372	0.455 ± 0.025	0.369	0.463 ± 0.026
f_2	0.119	0.124 ± 0.015	0.118	0.107 ± 0.014
f_3	0.020	0.033 ± 0.007	0.020	0.029 ± 0.007
f_4	0.001	0.002 ± 0.003	0.001	$0.007 \pm 0.004'$
f_5	—	0.002 ± 0.002	—	0.000 ± 0.002
f_6	—	0.001 ± 0.001	—	—
n	0.674	0.827 ± 0.032	0.669	0.792 ± 0.032
v	1.27	1.55 ± 0.06	1.26	1.49 ± 0.06
	Au		Pb	
f_0	0.466	0.368 ± 0.022	0.454	0.376 ± 0.027
f_1	0.373	0.447 ± 0.023	0.378	0.446 ± 0.028
f_2	0.135	0.144 ± 0.014	0.143	0.121 ± 0.016
f_3	0.024	0.027 ± 0.006	0.024	0.049 ± 0.010
f_4	0.002	0.011 ± 0.003	0.001	0.007 ± 0.004
f_5	—	0.002 ± 0.002	—	0.002 ± 0.002
f_6	—	—	—	—
n	0.722	0.870 ± 0.029	0.740	0.873 ± 0.037
v	1.36	1.63 ± 0.06	1.40	1.64 ± 0.07

Fig. 4. Comparison of the calculated [13] neutron multiplicities with the experiments of Kaplan et al. [1, 11] (assuming a Gaussian momentum distribution with $\alpha^2/2M = 20$ MeV and $M^* = 0.68\,M$)

heavy nuclei. A further improvement is obtained when the emission of neutrons from the nuclear surface due to the pseudodeuteron effect [16] is also included in the calculation. As can be seen by comparing Fig. 1 and Fig. 6, the agreement for Ag is now fair, for a Gaussian distribution with $\alpha^2/2M = 20$ MeV, $M^* = 0.39\,M$, direct volume emission of 22% and clustering parameter of 0.14.

Further to the early radiochemical methods [17] for identifying the radioactive products following muon capture and the direct measurements of the emitted neutrons of Kaplan et al. [11, 1] with a high-efficiency detector, a third generation of detailed experiments [18–22, 84–87] has recently been started in which the γ-rays following μ-capture are detected. Since in most cases one would expect the nucleus to be left in an excited state after emitting a neutron, or even when capture occurs without neutron emission, the multiplicity distribution of neutron-emission can be obtained from the identification of the isotopes detected through their γ-emission. These nuclear γ-rays are delayed with respect to the X-rays because of the finite lifetime of the muon in the atomic $1S$ level. The main disadvantage of

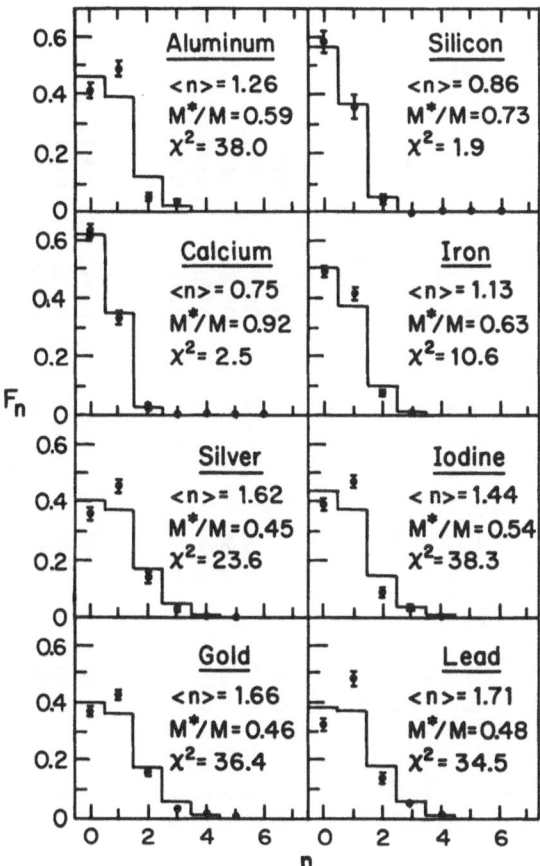

Fig. 5. Comparison of the calculated [1] neutron multiplicities with experiment [1], by using a Gaussian momentum distribution with $\alpha^2/2M = 20$ MeV and effective masses chosen as to give the experimental average multiplicity

this method is that one cannot detect neutron emission leading to the ground state of the residual nucleus. On the other hand, one can measure directly the non-neutron capture events by detecting the γ-rays from the excited $_{Z-1}A^{N+1}$ nucleus appearing after μ-capture. Moreover, this method allows of a direct measurement of the relative occurrence of various excited states appearing in the process.

Backenstoss et al. [18] have measured the nuclear γ-rays after muon capture in several odd-A, odd-Z isotopes — $^{45}_{21}Sc$, $^{55}_{25}Mn$, $^{59}_{27}Co$, $^{93}_{41}Nb$, $^{127}_{53}I$, and $^{209}_{83}Bi$, Petitjean et al. [19] in $^{151}_{63}Eu$ and $^{153}_{63}Eu$ and Kessler et al. [20] in $^{89}_{39}Y$. The comparison for the neutron multiplicities obtained in the first two experiments with calculations based on the

Target	Direct-emission parameter	Clustering parameter	Effective mass M^*/M	Multiplicity distribution						
				x^2	P_0	P_1	P_2	P_3	P_4	P_5
Fermi gas, $\theta_f = 0$ MeV										
Ag	0.216	0	0.43	4.7	0.343	0.463	0.166	0.029		
I	0.199	0	0.45	20.2	0.378	0.474	0.136	0.012		
Au	0.157	0	0.34	27.9	0.333	0.458	0.179	0.030		
Pb	0.153	0	0.32	15.2	0.320	0.459	0.190	0.030		
Ag	0.216	0.144	0.45	9.5	0.333	0.472	0.175	0.019		
Fermi gas, $\theta_f = 12$ MeV										
Ag	0.216	0	0.26	24.4	0.409	0.377	0.151	0.054	0.009	0.001
I	0.199	0	0.30	32.2	0.442	0.380	0.135	0.039	0.004	
Au	0.157	0	0.19	41.5	0.407	0.364	0.158	0.058	0.011	0.001
Pb	0.153	0	0.18	35.7	0.394	0.368	0.165	0.060	0.012	0.001
Ag	0.216	0.144	0.29	11.8	0.388	0.397	0.167	0.041	0.006	
Gaussian, $\alpha^2/2M = 20$ MeV										
Ag	0.216	0	0.36	22.3	0.407	0.380	0.149	0.053	0.009	0.001
I	0.199	0	0.46	29.6	0.441	0.384	0.133	0.038	0.004	
Au	0.157	0	0.40	32.7	0.401	0.373	0.160	0.056	0.010	0.001
Pb	0.153	0	0.41	30.4	0.387	0.377	0.167	0.057	0.011	0.001
Ag	0.216	0.144	0.39	10.5	0.387	0.400	0.166	0.041	0.006	

Fig. 6. Multiplicity distributions calculated [1] with the inclusion of direct volume and surface emission [13]

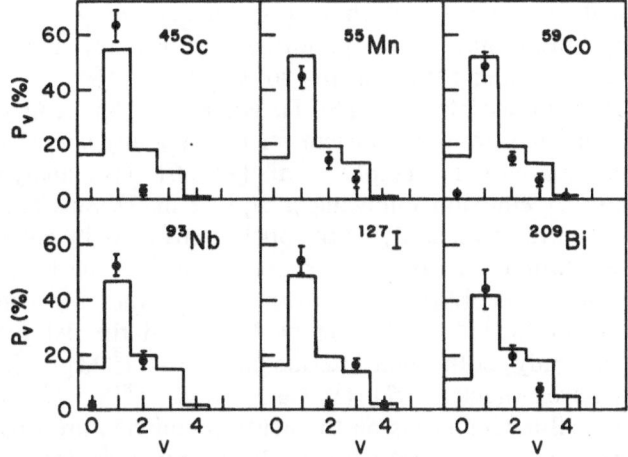

Fig. 7. The experimental values of the neutron multiplicities for several nuclei, found by γ-ray detection [18], compared with the predictions based on the model of Ref. [13]. Parameters used in the calculation: Gaussian momentum distribution with $\alpha^2/2M = 20$ MeV, $M^* = 0.5 M$, nuclear temperature $\theta = 0.75$ MeV

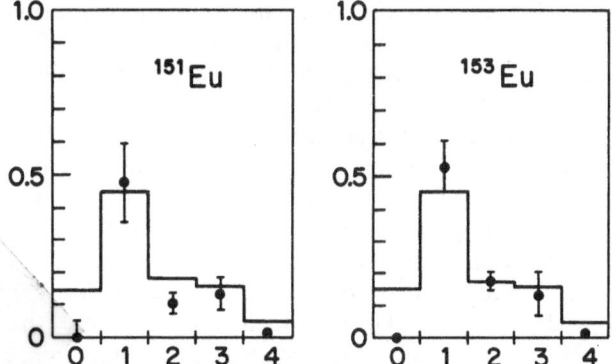

Fig. 8. The experimental values of the neutron multiplicities for $^{151}_{63}$Eu and $^{153}_{63}$Eu found by γ-ray detection [19], compared with the predictions based on the model of Ref. [13]. Parameters used in the calculation: Gaussian momentum distribution with $\alpha^2/2M = 20$ MeV, $M^* = 0.5 M$, nuclear temperature $\theta = 0.75$ MeV

model of Ref. [13] is given in Figs. 7 and 8. In the calculation the direct nucleon emission has been included (though not any surface effects [16]), along with the evaporation neutrons. These experiments indicate practically zero probability for no-neutron emission for all the nuclei studied, while the theoretical calculation predicts 15–18% for it. On the other side, the agreement with the calculated multiplicity distributions for emission of one or more neutrons is quite good.

Additional experiments which detect the delayed γ-ray emission following μ-capture have been performed recently by Evans [84], Temple et al. [85] and Miller et al. [86, 87]. Evans [84] has studied a large number of nuclei, including Si, Ti, Mn, Fe, Co, Ni, Y, Ag and Au, however his findings are not sufficient for drawing any conclusions on the neutron distribution. Temple et al. [85, 86] have analyzed the delayed γ-nuclear emission following μ-capture in Al, Si, Ca, and Co and have concentrated on studying the nuclear structure by comparison with other excitation reactions in appropriate nuclei. The lower limits they were able to establish on neutron multiplicities are consistent with the data of Ref. [1]. Miller et al. have used the delayed γ-ray technique to study both light-intermediate [86] ($^{24}_{12}$Mg, $^{28}_{14}$Si) and heavier even-even nuclei [87] ($^{142}_{58}$Ce, $^{140}_{58}$Ce, $^{138}_{56}$Ba, $^{120}_{50}$Sn). The sought after shell-model effects on the neutron emission are suggestive, but not definitive. The zero-neutron de-excitation is found to be vanishingly small ($< 2\%$) in the four heavy nuclei studied, in agreement with the general trend previously established in these type of experiments [18, 19].

In the experiments of MacDonald et al. [1] the probability for the formation of isotopes with the same mass as the capturing nucleus was found however to be high for intermediate nuclei (19% in $^{56}_{26}$Fe) and between 5–10% for heavy nuclei (except Pb for which they also found $< 1\%$). It should be remarked that in the γ-ray experiments [18–20, 87] only 60–80% of the capturing processes are being accounted for. Those unaccounted for are presumably capture events in which no excited nuclei are left. It is however hard to believe that all the μ-capture events without neutron emission go directly to the ground state of the resultant nucleus – which would account for the discrepancy. One should also add that recent activation experiments by Bunatyan et al. [23, 24], in which they, measure the relative probability $W_{\mu,\nu}$ for muon capture without subsequent neutron emission, tend to support the findings of MacDonald et al. [1] for intermediate nuclei. In particular, they find $W_{\mu,\nu}$ to be $16 \pm 3\%$ in $^{56}_{26}$Fe, $10 \pm 1\%$ in $^{27}_{13}$Al, $28 \pm 4\%$ in $^{28}_{14}$Si and $10 \pm 1\%$ in $^{51}_{23}$V, in fair agreement with the findings of Ref. [1].

On the other hand, activation experiments with heavy nuclei performed by the same group [88] give $W_{\mu,\nu}$ to be $9 \pm 1.5\%$ in ^{208}Pb and $4 \pm 1\%$ in $^{139}_{57}$La. At least the figure for ^{208}Pb(μ^-, ν) ^{208}Tl appears to be in disagreement with the $0 \pm 3\%$ result given by Kaplan et al. [1, 89] for no-neutron emission after μ-capture in natural lead.

Heusser and Kirsten [90] have recently used advanced techniques in low-level γ-ray spectrometry to perform essentially a new type of activation experiment in muon capture physics. In their experiment, performed on Fe, Ni, Co, Mg, and Al, the yield of the radioisotopes produced after μ-capture is determined by non-destructive γ-spectrometry, without any chemical separation, the radioisotopes being identified by both their characteristic γ-rays and half-lives. For capture in Ni they are able to determine quite accurately the neutron multiplicity distribution, since most of the μ-captures lead to detectable Co radioisotopes. Thus, they account for 80% of the captures in Ni, a figure comparable with the highest detected yields with the γ de-excitation method [18–20, 85–87]. Their result for capture in the various Ni isotopes (^{58}Ni (67.9%), ^{60}Ni (26.2%), ^{62}Ni (3.2%)) compared to the calculated yields by using the model of Ref. [13] (including direct emission), is given below:

Reaction	Measured yield (%)	Calculated yield (%)	
		$M^* = 0.5\,M$	$M^* = 0.8\,M$
^{58}Ni (μ^-, ν) ^{58}Co	24.3 $\pm$ 2.0	19.0	22.3
^{58}Ni $(\mu^-, \nu n)$ ^{57}Co	41.3 $\pm$ 2.9	39.0	41.3
^{58}Ni $(\mu^-, \nu 2n)$ ^{56}Co	4.6 $\pm$ 0.5	13.4	8.1
^{58}Ni $(\mu^-, \nu 3n)$ ^{55}Co	0.36 $\pm$ 0.11	4.2	0.48
^{60}Ni (μ^-, ν) ^{60}Co	7.9 $\pm$ 0.6	6.1	7.6
^{62}Ni $(\mu^-, \nu n)$ ^{61}Co	2.5 $\pm$ 0.5	2.2	2.5
Ni $(\mu^-,$ charged part.)	2.6 $\pm$ 0.3		
^{58}Ni $(n, 2n)$ ^{57}Ni	0.08 $\pm$ 0.02		
	$\Sigma = 80.7 \pm 3.8$		

The agreement with theory for $M^* = 0.8\,M$ is, indeed, remarkable for all the measured neutron multiplicities. The unaccounted yield is estimated to be due mainly ($\sim 16\%$) to capture leading to the stable ^{59}Co isotope [^{60}Ni (μ^-, n) ^{59}Co] and the rest of $\sim 4\%$ to captures leading to $^{62-64}$Co and unobserved charged particle emissions. Other interesting results of Heusser and Kirsten [90] are the yields for neutronless captures in ^{59}Co (μ^-, ν) ^{59}Fe $(15.1 \pm 1.1\%)$, ^{56}Fe (μ^-, ν) ^{56}Mn $(20.1 \pm 1.3\%)$ and ^{24}Mg (μ^-, ν) ^{24}Na $(17.0 \pm 2.6\%)$. The latter figure appears consistent with the findings of Miller et al. [86], and the result for ^{56}Fe agrees well with measurements done with different techniques [1, 23]. On the other hand, Backenstoss et al. [18] give for neutronless capture in ^{59}Co a result which is 8 times (!) smaller than that of Heusser and Kirsten. The theoretical yields for neutronless capture calculated with Singer's model [13] for $M^* = 0.5\,M$ agree very well with the results of Heusser and Kirsten, giving 15.4% for ^{59}Co, 17.1% for ^{56}Fe and 19.2% for ^{24}Mg.

Of special interest is the role played by the giant multipole resonances in the muon capture [6–8]. One can expect that in particular in light and intermediate nuclei, the deexcitation of the giant excited states would manifest itself in line spectra of emitted neutrons. Calculations of these spectra for several light nuclei have been performed by Überall et al. [25, 26, 27] and recently experiments have reported the observation of the predicted lines superimposed on the evaporation spectrum. Evseev et al. [9, 28] reported this effect for ^{16}O, ^{32}S, and ^{40}Ca and Plett and Sobottka [29] for ^{12}C and ^{16}O.

Temple et al. [85] probe the resonance μ-capture mechanism by comparing the relative probability of exciting various states in the resultant nucleus, with the state populations obtained in other reactions like photo and electro-excitation. Thus, they compare ^{40}Ca $(\mu^-, \nu n\gamma)$ ^{39}K with ^{40}Ca$(\gamma, p\gamma')$ ^{39}K and they find that, although the same states are excited in both reactions, the correlation between the state populations is not very strong. Still, the results can be considered as evidence for the presence of the giant resonance mechanism in the μ-capture in ^{40}Ca. Likewise, they find evidence for a giant M 1 resonance mechanism in the μ-capture process ^{28}Si $(\mu^-, \nu\gamma)$ ^{28}Al. Miller et al. [86] find good correlation for exciting the same levels in Al^{27} by the reactions ^{28}Si $(\mu^-, \nu n)$ ^{27}Al and ^{28}Si (γ, p) ^{27}Al, thus providing another evidence for a giant resonance mechanism in the μ-capture process. This topic is being analyzed in detail in Prof. Überall's talk at this Conference [81].

As the experiment indicates that the majority of the emitted neutrons in intermediate nuclei (e.g. Ca^{40}) are nevertheless of the evaporation type [5], it is of interest to calculate their emission when the excitation nuclear function is calculated by taking into account the transitions to the giant resonance states. Singer and Zin [30] have obtained the excitation function of μ-capture in Ca^{40} by using the approach of Foldy and Walecka [7] of relating the contribution to the capture probability of the giant dipole states to the appropriate photo-absorption cross section. One obtains a nuclear excitation function

$$I(Q) = A(E_m - Q)^4 \, \sigma_\gamma(Q)/Q + B(E_m - Q)^6 \,, \qquad (9)$$

A, B being constants determined by the relative contribution of the dipole and the higher states to the capture probability. The multiplicity distributions calculated [30] from (9) are in good agreement with experiment [1].

2.3 Direct Neutrons and Neutron Energy Spectrum

A smaller part of the neutron emission occurs in a "direct" manner. Namely, the neutron formed in the weak process (1) succeeds in escaping from the capturing nucleus, usually with a higher energy than the boil-off neutrons evaporated from the intermediate compound state. The direct neutrons bear information on the capture process, and the study of the spectrum of the emitted neutrons can also be a helpful tool in the understanding of the nuclear physics involved in the process. It is therefore of interest to have good data on the shape of the energy spectra of the emitted neutrons.

The earliest detailed experiment is due to Hagge [31], who measured the neutron energy spectrum between 2 and 16 MeV in calcium and lead. In this experiment the energy spectrum was measured by accumulating the pulse height spectrum resulting from neutron detection in a liquid scintillator. The spectra seem to contain a fair amount of evaporation neutrons in the low energy region. No separation or normalization to the total number of neutrons expected is made. Turner [32] has measured the energy spectra in the same nuclei in a hydrogen bubble chamber experiment. Energies of protons from neutron-proton scatterings are obtained from range relations, and then the neutron spectrum is deduced. Although neutrons with energies as high as 50 MeV are detected, only the part of the spectrum up to about 20 MeV has reasonable statistics. It appears that beyond 5 MeV the evaporation spectrum is negligible, and the percentage of

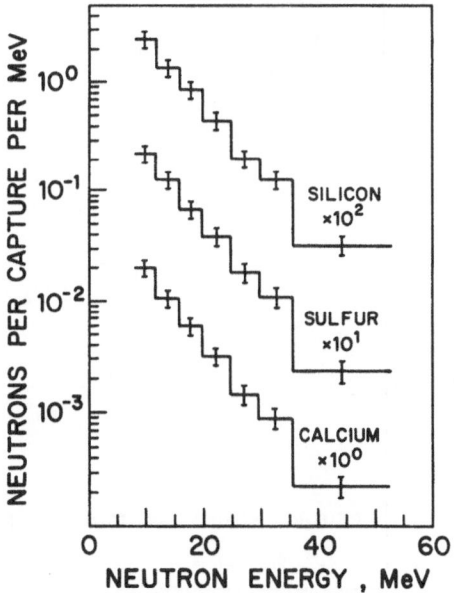

Fig. 9. Energy spectra of high energy neutrons emitted after muon capture in Si, S and Ca (Ref. [3]). These data remain essentially unchanged in the second experiment of Edelstein and Sundelin (Ref. [33])

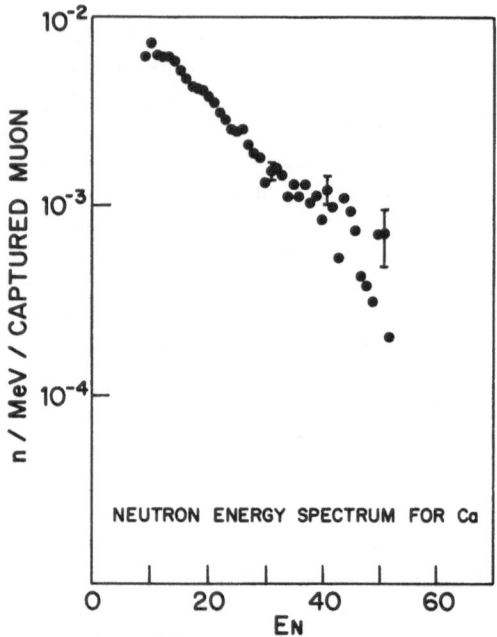

Fig. 10. Energy spectrum of high energy neutrons emitted after muon capture in Ca. (Ref. [4])

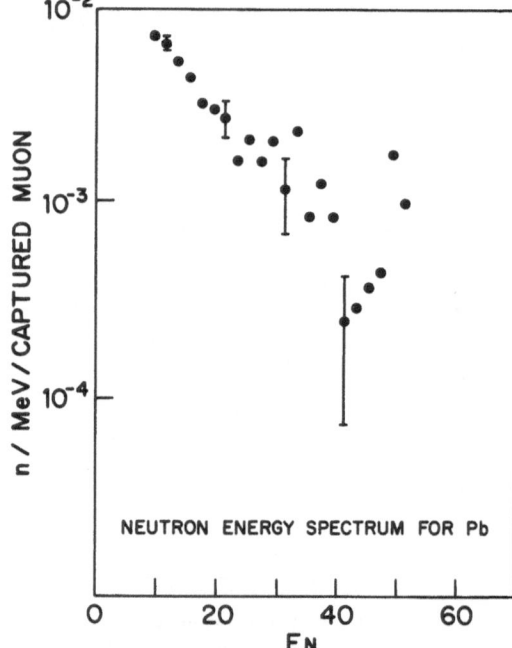

Fig. 11. Energy spectrum of high energy neutrons emitted after muon capture in Pb (Ref. [4])

directly emitted neutrons is estimated to be $20 \pm 2\%$ in calcium and $12 \pm 2\%$ in lead.

Better statistics experiments were performed recently by Krieger [4], who uses a spark chamber-scintillator neutron detector, which gives a measurement of neutron energy through the measurement of the recoiling proton in the spark chamber, and by Sundelin et al. [3, 33] who obtain the neutron spectrum by unfolding the observed pulsed-height proton spectrum. Sundelin et al. measure the spectrum in Si, S, and Ca between 7.7–52.5 MeV (Fig. 9), while Krieger gives the resulting spectrum from capture in S, Ca (Fig. 10) and Pb (Fig. 11) between 10 and 40 MeV. In both experiments one finds that the high energy neutron spectrum falls approximately in an exponential manner:

$$N(E) \sim \exp\left(-E/E_0\right). \tag{10}$$

The values obtained for the exponential constant in the two experiments are given below:

	Ref. [3]	Ref. [4]
Si	7.6	
S	7.3	15 ± 1
Ca	7.2	14 ± 1
Pb		12 ± 1

Experimental values for the constant E_0 (in MeV) of Eq. (10)

Although the experiments agree on the general shape, there is clear disagreement on the value of the constant E_0.

From the experiments of Krieger [4] and Sundelin et al. [3, 33] one has also information on the percentage of direct neutrons emitted per muon capture process. Krieger [4] gives for the percentage of neutrons emitted with energies higher than 10 MeV:

$E_n > 10$ MeV   for S 0.159 ± 0.008 neutrons/μ capture

for Ca 0.102 ± 0.005 neutrons/μ capture

for Pb 0.091 ± 0.006 neutrons/μ capture.

The comparable figures from the experiment of Sundelin et al. [3, 33] are as follows:

$E_n > 7.73$ MeV   for Si 0.247 ± 0.044 neutrons/μ capture

for S 0.210 ± 0.037 neutrons/μ capture

for Ca 0.179 ± 0.031 neutrons/μ capture

$E_n > 11.49\,\text{MeV}$ for Si 0.139 ± 0.025 neutrons/μ capture

for S 0.122 ± 0.022 neutrons/μ capture

for Ca 0.102 ± 0.018 neutrons/μ capture

In two other recent experiments measurements have been performed in the energy range of 1–15 MeV, thus covering the evaporation part as well as a portion of the high-energy part of the spectrum. Evseev et al. [9, 28] used a stilbene crystal to detect protons, and the neutron spectra from μ-capture in ^{16}O, ^{32}S, ^{40}Ca, and Pb were obtained by differentiation of proton spectra. Their results are given in Fig. 12.

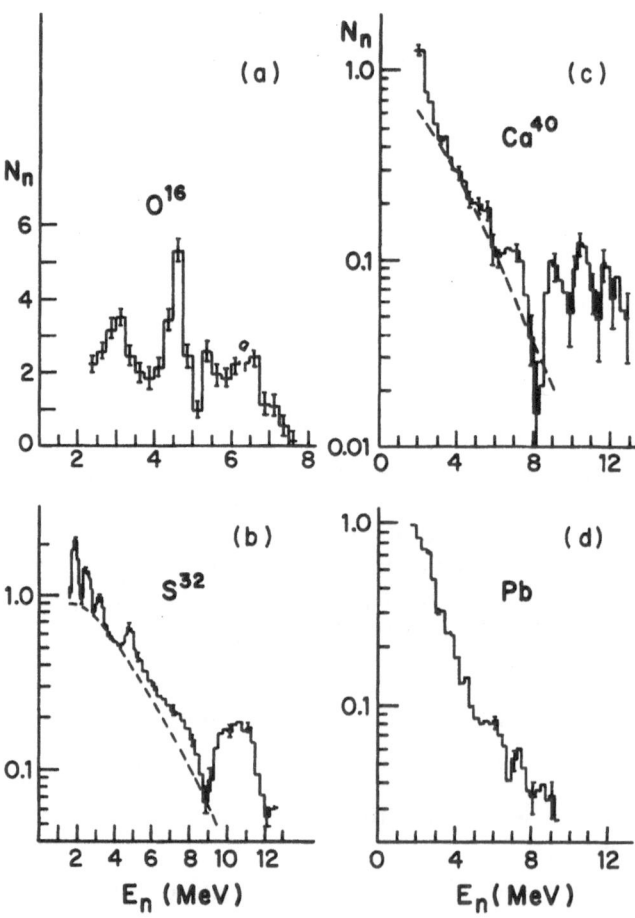

Fig. 12. Neutron energy spectra from muon capture in ^{16}O, ^{32}S, ^{40}Ca and Pb (Ref. [9])

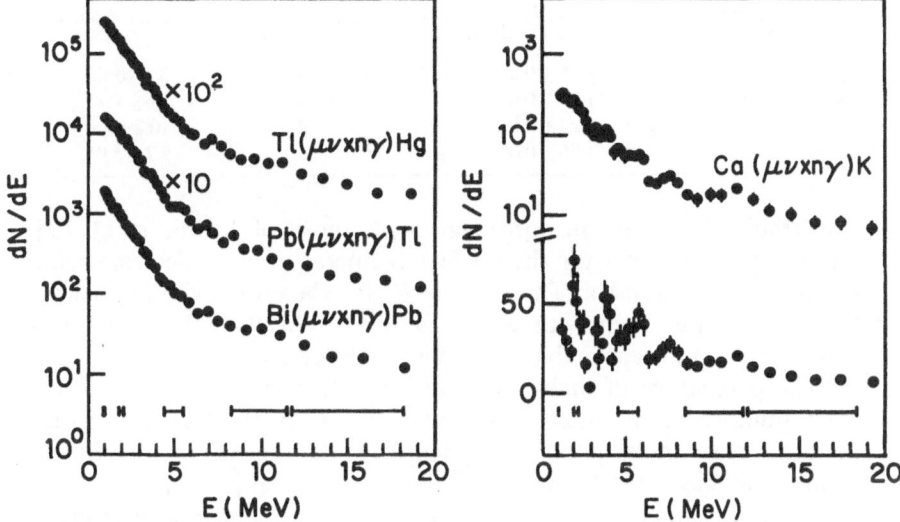

Fig. 13. Spectra of neutrons from muon capture in $_{81}$Tl, $_{82}$Pb, $_{83}$Bi and $_{20}$Ca (Ref. [5]). For capture in Ca, the total spectrum is given at the top, and that obtained after subtracting an evaporation spectrum, at the bottom

From this experiment one has evidence of line spectra in light nuclei, and characteristic lines are observed also in ^{40}Ca. The main part of the spectra in the low energy region is of evaporational type and the characteristic temperatures of the final nuclei formed after capture in S, Ca, and Pb are

$$(\theta)_S = 1.7 \,\text{MeV}, \qquad (\theta)_{Ca} = 1.5 \,\text{MeV}, \qquad (\theta)_{Pb} = 1.15 \,\text{MeV} . \tag{11}$$

Schröder et al. [5] measure for the first time the spectra of neutrons from μ-capture in coincidence with de-excitation γ-rays from the resultant nucleus. The neutron energy was determined directly by using a time-of-flight method, the γ-ray serving as a signal to start the neutron measurement. The results of their measurements on neutrons in the energy range 1–20 MeV from μ-capture in Ca, Tl, Pb, and Bi are given in Fig. 13. The spectra are fitted with an evaporation-type spectrum of the form

$$dN(E) \sim E^{5/11} \, e^{-E/\theta} \, dE \tag{12}$$

in the energy range $E < 4.5\,\text{MeV}$, and with an exponentially decaying form [Eq. (10)] for energies $E > 4.5\,\text{MeV}$. Their results for the nuclear temperature θ, the high-energy decay constant E_0, and the percentage I of neutrons with energies larger than 4.5 MeV are given below:

Target element	$\theta(\text{MeV})$	$E_0(\text{MeV})$	$I(E > 4.5\,\text{MeV})$
$^{40}_{20}\text{Ca}$	1.35 ± 0.07	8.4 ± 1.0	$3 - 20\%$
$^{203,\,205}_{81}\text{Tl}$	1.09 ± 0.04	8.8 ± 1.1	$9.6 \pm 0.8\%$
$^{206,\,207,\,208}_{82}\text{Pb}$	1.22 ± 0.06	9.0 ± 1.2	$10.2 \pm 1.0\%$
$^{209}_{83}\text{Bi}$	1.06 ± 0.05	8.2 ± 1.2	$9.7 \pm 1.0\%$

The results for θ are in agreement with those of Evseev et al. [9] and the general trend for the nuclear temperature is to decrease with increasing mass number. The value of E_0 for Ca agrees with Sundelin's [3] much better than with Krieger's result. On the other hand the value for Pb is in reasonable agreement with Krieger's.

The percentage of high-energy neutrons decreases with increasing mass number in all these experiments [3, 4, 5, 9]. On the whole, there is reasonably good agreement on the percentage of direct neutrons as measured by Krieger [47] and Sundelin [3, 33], compared to the results of Schröder et al. [5]. The latter found approximately 10% of direct neutrons with energies higher than 4.5 MeV for all the heavy nuclei studied, while Krieger detected $\sim 5.5\%$ of high energy neutrons with $E > 10$ MeV in Pb (normalizing to 1.7 neutrons per μ-capture). In view of the complex structure detected [5] in Calcium, a comparison with the data of Krieger and Sundelin is rather ambiguous, but it appears that at least Sundelin's figures [3, 33] for the percentage of high energy neutrons are higher than those reported in Ref. [5].

It should be remarked, however, that a comparison of the results of Schröder et al. [5] with those of other experiments [3, 4, 9, 33] has a serious limitation in the fact that the latter are able to record absolute values for the amount of neutrons detected in a certain energy range (normalized per muon capture), while the values of Ref. [5] are expressed with relation to their detected spectrum. Nevertheless, it is very interesting that, although in Ref. [5] only those neutrons which originate from excited states are detected (in coincidence with γ's), the energy spectrum is quite similar to that of the other experiments in which also transitions to ground states are detected. Thus, the above mentioned comparison is probably more meaningful than one might have expected.

Before proceeding to discuss theoretical work, it is interesting to remark that even in light nuclei there is a sizable amount of evaporation neutrons emitted after μ-capture. Plett and Sobottka [29] have found 62% of the emitted neutrons after capture in O^{16} to fit an evaporation spectrum with $\theta = 1.25$ MeV, while for C^{12} the appropriate figure is 38%, with a nuclear temperature $\theta = 1.01 \pm 0.30$ MeV.

Theoretical work on the mechanism for direct neutron emission following muon capture in intermediate and heavy nuclei has had only limited success in accounting for the observed spectrum of high energy neutrons or for the percentage of direct emission. In fact, the early calculations predicted the vanishing of the high energy neutron spectrum beyond 15–20 MeV.

Überall [34] calculated the spectrum of the directly emitted neutrons for a nucleus described by a Fermi gas model with equal numbers of protons and neutrons. Absorption or scattering of the neutrons in the nucleus is neglected. The spectrum obtained decreases linearly with energy and vanishes beyond 16 MeV. Lubkin [35] performed a "modified Fermi gas" calculation for Ca^{40} and Pb^{208}, dividing the nucleus into small boxes and treating the nucleons in each box with the Fermi gas model. The μ-capture neutron can thus be considered to appear in a localized zone of the nucleus. The thus incoherently-produced neutrons are subsequently followed out of the nucleus by geometrical optics, allowing for attenuation in the nuclear volume, and refraction and total reflection at the nuclear surface. The spectrum calculated for Ca has a peak around 5 MeV and vanishes beyond 10.77 MeV, while that for Pb peaks at 2 MeV and vanishes beyond 4.56 MeV. Dolinsky et al. [36] used the shell-model with $j-j$ coupling for Ca^{40}, and the interaction of the outgoing neutron with the nucleus is treated by using an optical potential. The spectrum obtaines peaks also around 5 MeV, and then falls approximately exponentially to zero beyond 20 MeV. The spectra obtained in Refs. [34–36] are given in Fig. 14.

Singer [13] used a simple model to roughly estimate the probability of direct emission of neutrons from the nuclear volume following μ-capture. The nucleus is taken as a constant-density sphere and the probability for direct escape p is calculated from

$$p = P \times d \qquad (13)$$

where d is the average probability of a neutron created at some point in the nucleus reaching the boundary, and P the probability of its crossing the centrifugal barrier. Using

$$d = \exp\left[-0.75\, R/\lambda\right], \qquad (14)$$

where λ is the mean free path in nuclear matter and

$$P(E_n) = \Sigma\, B_l P_l(E_n), \qquad (15)$$

where B_l is the probability of occurrence of l, and $P_l(E_n)$ the escape probability of a neutron of angular momentum $l\hbar$ and energy E_n,

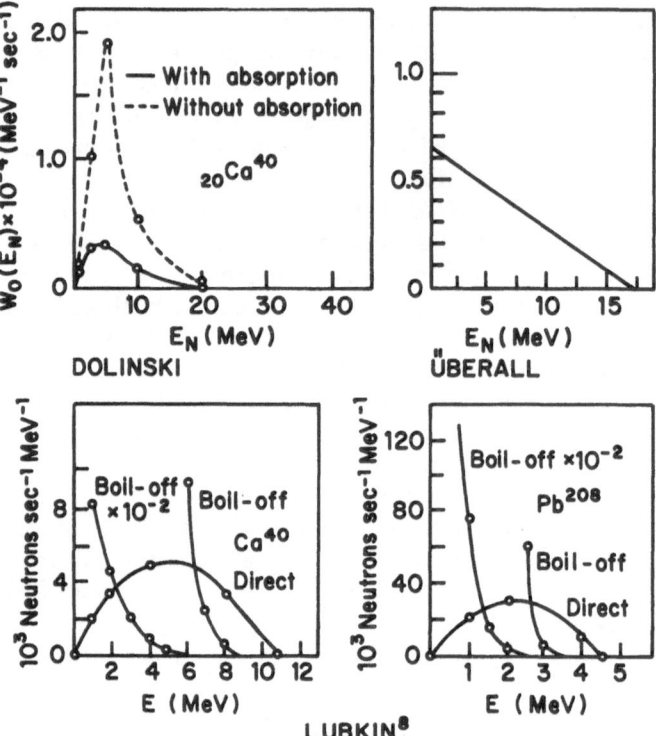

Fig. 14. Spectra for direct neutrons as calculated by Dolinsky et al. [36], Überall [34] and Lubkin [35]

one finds [13] for the percentage of direct emission in several heavy nuclei

Ag: 22%,   I: 20%,   Au: 16%,   Pb: 15%   (16)

The overall dependence on the mass number obtains correctly in this simple model. The calculation for Pb gives 9% direct neutrons out of a total average emission of 1.7 neutrons per μ-capture, in excellent agreement with Schröder's findings [5], as well as with those of Krieger [4] [5.5% direct neutrons with $E_n > 10$ MeV and remembering that the calculated figures in (16) refer to the whole spectrum of direct neutrons, i.e. $E_n \gtrsim 4$ MeV]. The constancy of the percentage of neutrons emitted directly for the three heavy elements studied by Schröder et al. [5] ($\sim 10\%$) seems to confirm the physical picture of the model of Ref. [13]. It would be interesting to have similar measurements for nuclei with $A = 100 - 120$, so as to check whether this picture still holds also for this region [from (16) one expects 15%

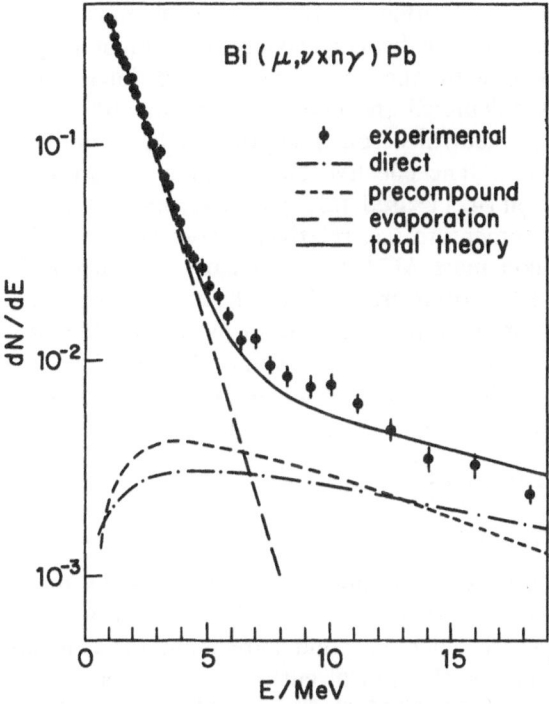

Fig. 14a. Experimental and theoretical neutron spectrum from μ-capture in Bi (Ref. [5]). The experimental spectrum is normalized to fit the theoretical one

direct neutrons out of the total average emission of ~ 1.5 per capture in this region]. The model has been refined by Schröder [5], who used it to calculate the combined evaporation and direct emission neutron energy spectrum for capture in Tl, Pb, and Bi. The agreement with the experimental spectrum of Ref. [5] is impressive, as it can be seen from Fig. 14a for capture in Bi. The direct spectrum is calculated by using a square well optical potential when following the neutron's behaviour within the nucleus. In addition, the "pre-compound" emission is also included by considering separately the neutrons which have been scattered once on other bound nucleons. The calculated integrated intensity of the direct and "pre-compound" emission amounts to 14% per capture. By adding to it a neutron evaporation spectrum for the 86% of cases of compound nucleus formation, the solid line of Fig. 14a is obtained.

Recent theoretical calculations [37, 38], which take into account relativistic effects in muon capture [39–41] as well as the interaction between the directly emitted neutrons and the nucleus, succeed in obtaining a much more realistic spectrum of the high-energy neutrons, in fair agreement with the measurements in intermediate nuclei. Bogan [37] has calculated the energy spectrum and the asymmetry of the high-energy neutrons from μ-capture in Si, S, and Ca with the following model: (a) The effective Hamiltonian for muon capture of Fujii and Primakoff [42] is used, taking into account terms proportional to the nucleon momentum (i.e. relativistic corrections to the order of the inverse nucleon mass M^{-1} are included); (b) Harmonic-oscillator shell-model wave functions are used for the capturing proton with the harmonic-oscillator parameter chosen so as to give the proper r.m.s. charge radius; (c) Plane-waves are assumed for the emitted neutrons, their interaction with the residual nucleus being taken into account by using for the neutron momentum inside the nucleus

$$p_n^2 = 2m^* m_p (E_N + B_W) + P_F^2, \tag{17}$$

where $m^* = 0.6 \, M_N$ is the neutron effective mass, B_W the binding energy of the most weakly bound proton, and P_F the nuclear Fermi momentum. His results are compared with the experiment of Sundelin et al. [3] in Fig. 15. There is good agreement for neutron energies above 25 MeV, but the theoretical curves are a factor 2–3 lower than this experiment around 10 MeV. It should be remarked that the improved agreement with experiment is due mainly to the nuclear wave functions used and the inclusion of the final state interaction through (17). Piketty and Procureur [38] have shown by an exact treatment of the Fermi-gas model that inclusion of relativistic terms to order M^{-2} does not affect the neutron spectrum (although they do contribute appreciably to the neutron asymmetry). Moreover, they have also calculated the high energy neutron spectra from μ-capture in ^{12}C, ^{28}Si, ^{32}S, ^{40}Ca by using also realistic shell-model wave functions for the capturing proton and eikonal-type distorted wave functions for the emitted neutron in order to take into account the final state interaction. The eikonal approximation is used for neutrons with energies $E_n \gtrsim 25$ MeV, with optical potential parameters deduced from neutron-nucleus scattering. They conclude that the final-state interaction has a strong effect on the neutron spectrum (Fig. 16).

This conclusion is confirmed by the calculations of Bouyssy and Vinh Mau [82, 91] who are able to obtain fair agreement [91] with the neutron spectrum of Sundelin and Edelstein [3, 33] by using Hartree-Fock wave functions and an optical potential whose imaginary part has both volume and surface components. Madurga [92] has also obtained a good spectrum in a related calculation (see Fig. 16a).

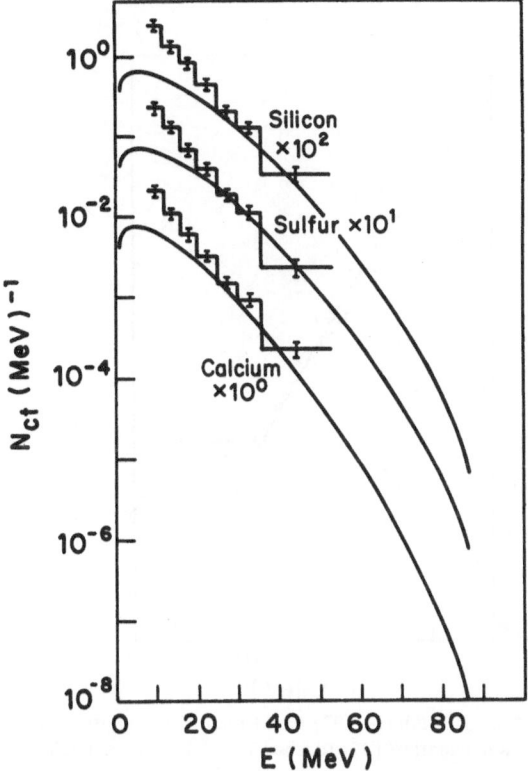

Fig. 15. Direct neutrons energy spectrum. The smooth curves are the calculation of Bogan (Ref. [37]) and the histograms are the experimental results of Sundelin et al. (Ref. [3])

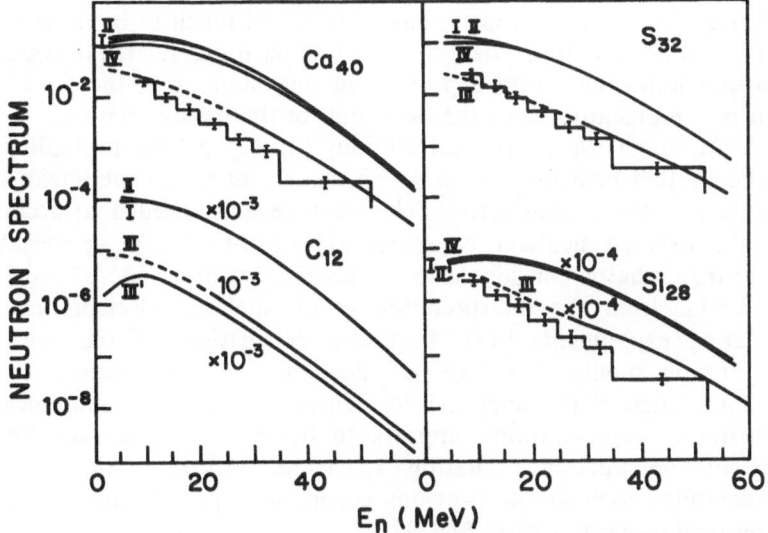

Fig. 16. Comparison of calculated neutron energy spectra by Piketty and Procureur (Ref. [38]) with the experimental results of Sundelin et al. (Ref. [3]) for Ca, S and Si

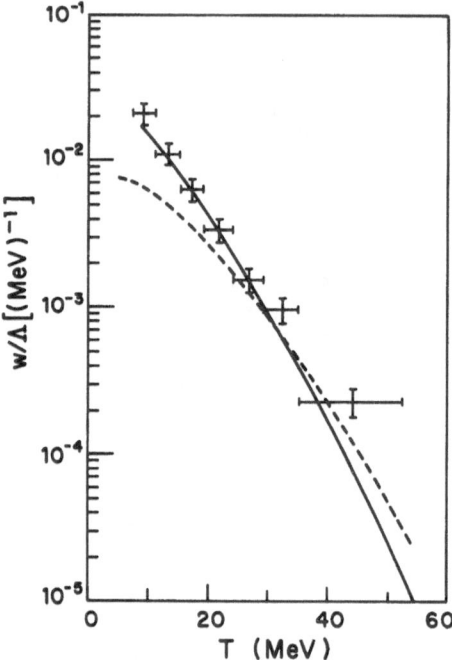

Fig. 16a. The calculation of Madurga [92] (solid line) of the neutron spectrum for Ca, compared with Bogan's calculation [37] (dashed line) and with Sundelin's data [3]

2.4 Remarks

1. From the above it appears that there is still much to be done both experimentally and theoretically on the various topics related to neutron emission following muon capture, and one hopes that the field will gain new momentum with the operation of the meson factories.

2. Although the simple calculations [1, 13] on the multiplicities of the emitted neutrons are in rough agreement with experiment, the picture is still not satisfactory. The effective mass needed to account for the average neutron emission, $M^* \simeq 0.4$–$0.5\,M$, is appreciably lower than what is considered to be the effective mass, i.e. $M^* \simeq 0.7\,M$ for $k < k_F$, from the interpretation of quasi-elastic electron-nucleus scattering experiments [43]. Improved calculations of the neutron evaporation should also take into account recent advances [44] in the knowledge of the nuclear-level densities, as the constant nuclear temperature representation appears to be valid only for the lower part of the nuclear excitation spectrum. Moreover the nuclear temperatures used in the previous calculations [1, 13] are generally lower than recently observed [5, 9].

3. There are very few attempts to account for the direct-neutrons spectrum in heavy nuclei. A calculation of the energy spectrum of direct and evaporation neutrons has been performed successfully by Schröder [5], using the simple model of Ref. [13], for the Tl, Pb, and Bi nuclei. With the advent of new experimental information, which has been accumulated in the last few years, it is high time for theorists to attack these problems, taking into account the details of nuclear structure. One would also like to have better theoretical estimates on the sensitivity of the emission processes to the nuclear models used.

4. On the experimental side, we are only at the beginning of experiments on the energy spectrum of the emitted neutrons, and there are some discrepancies between the existing experiments concerning the percentage of directly emitted neutrons. It will be of great value to have more experiments measuring *both the energy spectrum and the asymmetry of the emitted neutrons*, as well as experiments measuring *at the same time the multiplicities and the energy spectrum of the neutrons.*

5. There seems to be disagreement at present on the amount of mu-capture without neutron emission between the experiments measuring γ-rays [18, 19] and those measuring neutrons [1] or using activation methods [23, 24, 90].

III. Charged Particle Emission

3.1 General

The deexcitation of the nucleus by charged particle emission after muon capture is a very infrequent process in intermediate and heavy nuclei. Although very little experimental work has been done in this field (and likewise, hardly a matching amount of theoretical work), there is agreement on the scarcity of charged events between the pioneering extensive emulsion experiment of Morinaga and Fry [45] and the later experiments [46–48] of this type.

Morinaga and Fry [45] studied 24000 meson tracks which stopped in their emulsion. They found that the meson capture in the heavy element of the emulsion (Ag, Br) is accompanied in 2.2% of the cases by one proton emission and in 0.5% by alpha particle emission. The energy distributions of these charged particles were also obtained in the experiment and are given in Fig. 17. In a subsequent emulsion experiment [50], with nearly 4 times more statistics, Kotelchuk and Tyler [51] measured a proton energy spectrum in fair agreement with Morinaga and Fry (Fig. 18). It should be remarked that the proton spectrum of Morinaga and Fry [45] is not very accurate for energies

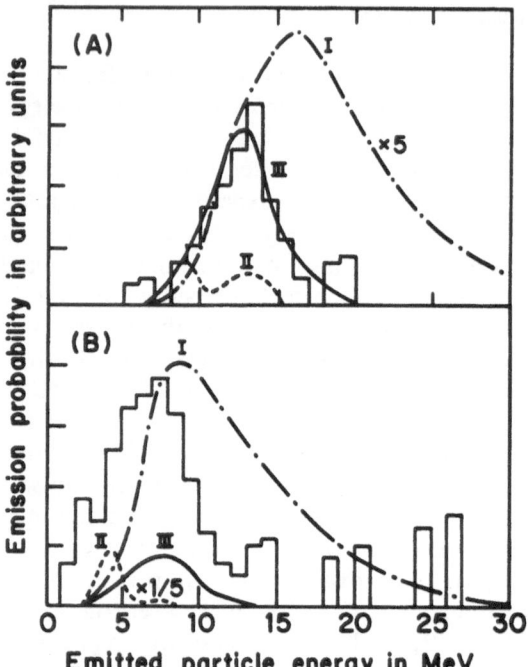

Fig. 17. Energy distribution of emitted alpha particles (A) and protons (B) following muon absorption in AgBr from the experiment of Ref. [45]. The curves are theoretical calculations of Ishii [49]

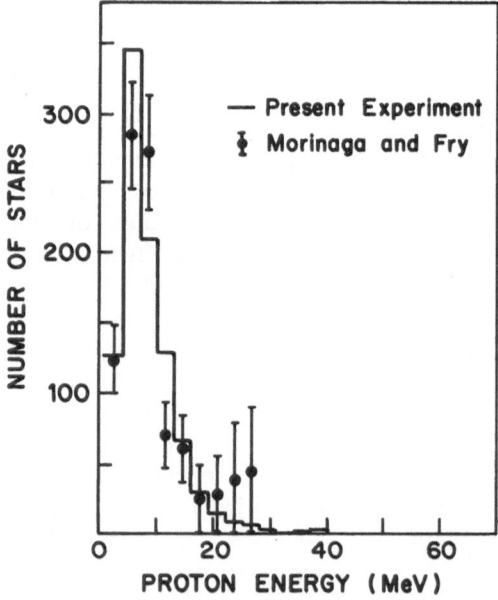

Fig. 18. Energy distribution of protons following muon capture in AgBr from the experiment of Ref. [51], compared with the earlier results of Morinaga and Fry [45]

> 15 MeV because of limited statistics and the need for large geometrical corrections.

Ishii [49] calculated the emission of protons and α-particles, assuming the formation of a compound nucleus following the μ-capture and subsequent statistical emission of various particles. He used for the nuclear level densities parameters determined from nuclear reactions at appropriate excitation energies. For the momentum distribution of protons in the nucleus he used three possible densities: (a) Fermi gas at zero temperature; (b) Fermi gas at a finite temperature ($kT = 9$ MeV) (c) The Chew-Goldberger distribution $\varrho(p) \sim A/(B^2 + p^2)^2$, which is known however from independent experiments to contain an inadequately high proportion of high momenta. Very good agreement with experiment for the α-emission is obtained with distribution (b), for both the absolute percentage (0.45% versus the experimental 0.5%) and the energy distribution (see curve *III* in Fig. 17(A)). The calculated emission of protons for the same finite temperature Fermi gas distribution turns out to be 10 times smaller than experiment. It is relevant that the distribution which accounts well for α-emission is quite similar to the one giving fair agreement [1, 13] with the experimentally observed neutron evaporation. Ishii also finds that (the unlikely) distribution (c) could account for the proton emission; however then the calculated α-emission is 15 times larger than observed.

In interpreting the results of Ishii's calculations one should also remember that at similar excitation energies the statistical theory explains satisfactorily α-emission in nuclear processes. On the other hand, under similar conditions the proton emission is frequently one to two orders of magnitude higher than that calculated from a compound nucleus with statistical emission [52, 53].

Assuming a reduced effective mass for the muon-absorbing nucleon the average calculated excitation energy available in the compound state will be increased. However, as charged particle emission takes place primarily from the momentum region near the Fermi level, and the effective-mass in this region is $M^*(p) \simeq M$, the inclusion of the effective-mass approximation does not significantly increase the calculated rate of proton emission. It appears therefore that the rate of proton emission following muon capture in Ag, Br significantly exceeds the number predicted by evaporation theory. Moreover, the observation of high energy protons with energies of 25–50 MeV also cannot be reconciled with the evaporation mechanism.

In view of the inability to account for the emission of protons as a statistical process, other mechanisms involving "direct" emission have been considered. Two models have been developed, which take into account the two-nucleon correlation, on the surface [16] as well as in the nuclear volume [54]. They are presented in the next sections.

3.2 Surface Correlation Effects (Pseudodeuteron Model)

There is experimental evidence [55] from various nuclear phenomena (including K^--capture), that the nuclear surface is relatively rich in nucleon clusters. Theoretical calculations [56, 57] account for the clustering tendency near the surface, from the behavior as a function of distance from the center of the nucleus of the correlation-free potential energy density versus the part of it that includes correlations.

Singer [16] has suggested that capture by two-nucleon clusters in the surface region can provide the mechanism for the emission of protons unaccounted for by the statistical calculations [49]. The elementary capture process in this case would be

$$\mu^- + 2N \to 2N' + \nu_\mu. \tag{18}$$

Capture by a two-proton cluster then results in the appearance of an energetic neutron-proton pair, with a fair chance of direct escape from the nucleus. This type of process will also contribute to direct one and two-neutron emission when the capture occurs on neutron-proton pairs and it has been shown [1, 13] that it does indeed improve the agreement with experiment of calculated neutron emission.

A pseudodeuteron-model calculation has been made by Singer [16] for Ag, Br nuclei, for which experimental findings [45–48] and statistical emission calculations [49] are available. It is assumed that the effective nucleon clusters are those in the classical forbidden region, which extends beyond the radius R_c at which the sum of the nuclear and Coulomb potential vanish. Using the charge distribution in nuclei obtained from high-energy electron-nucleus scattering, one has on the average 1.5 protons in that region in Ag, Br. Only pseudodeuterons in S-states are considered, which limits the proton-proton cluster to singlet-spin state. The capture probability from a proton cluster in Ag is then shown to be related [16] to the capture probability from a deuteron [58] ω_D by

$$\omega_{[p-p]0} = 1.74\,\omega_D \, \frac{|\phi_\mu, \mathrm{Ag}(R_c)|^2}{|\phi_{\mu,D}(0)|^2} \, \frac{2\pi\,(1 - \bar{\alpha} r_0)}{\bar{\alpha}\langle \alpha^2 + k^2 \rangle v}. \tag{19}$$

The differences between ω_D and $\omega_{[p-p]0}$ are due to: (1) different spatial wave functions, which can be compensated for by taking into account the ratio of normalization constants of the two wave functions [last factor in (19)]; (2) a different value for the muon wave function in the capture region; (3) the allowed final states [included in the numerical factor in (19)]. Assuming that half the protons produced after muon capture by a pseudodeuteron leave the nucleus, and that S-wave singlet proton-pairs in the nuclear surface behave 40% of the time [59] as pseudodeuterons, Singer finds [16] that 0.022 protons are emitted per μ-capture in Ag and Br. Keeping in mind that some of the figures used are uncertain (like the 40% clustering probability, the 50% probability for escape, the exact form of the pseudodeuteron wave function), the perfect agreement with the experiment of Morinaga and Fry [45] is to some extent fortuituous.

In the above calculation no energy spectrum for the emitted protons is deduced. The kinematics of the reaction however allows for the appearance of energetic proton from pseudodeuteron capture with energies up to $E_{max} \simeq 50$ MeV. Such energetic protons, unaccounted for by evaporation, have indeed been observed [48, 51]. The peak of the spectrum on the other hand is expected to be below 10 MeV.

3.3 Volume Correlation Effects

Accepting the result of Ishii [49] that the evaporated protons following muon capture in Ag and Br can account for only 10% of the proton emission, one has still to check whether some form of direct emission from the nuclear volume might possibly account for the observed rate. One can think of a direct process initiated by the neutron created in the weak-interaction process (1). The neutron could possibly knock out of the nucleus a proton on which it scatters. The direct "knock-on" mechanism for (p, p') reactions has been studied for nuclear reactions, where the inelastic scattering cross sections for protons with a few tenths of MeV are larger by order of magnitude than calculated by statistical process. Elton and Gomes [60] have shown however that the total reflection at the nuclear boundary essentially prevents these protons from directly leaking outside and the contemplated "knock-on" direct emission from the nuclear volume is even smaller than the compound nucleus emission. Hence the direct emission of protons from the nuclear volume following μ-capture appears to be very improbable in view of the results of Elton and Gomes [60]. It should be added that the large cross-sections for (p, p') reactions are accounted for by considering quasi-elastic scatterings taking place mainly in the nuclear surface [53, 60, 61].

Bertero, Passatore, and Viano [54] have considered another possibility for direct emission from the nuclear volume, taking into account nuclear correlations. In particular, they consider the possibility

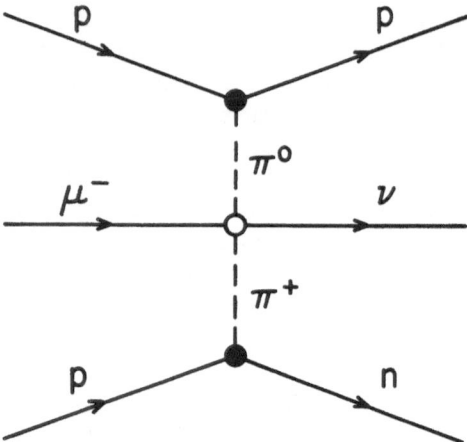

Fig. 19. The Feynman diagram for μ-capture by proton pairs considered by Bertero, Passatore and Viano [54]

of μ-capture by proton pairs, the weak ($\mu\nu$)-current interacting directly with the pion current exchanged by the nucleon pair. The Feynman diagram considered in their paper is exhibited in Fig. 19. The strength of the weak vertex involved is known (up to a form factor) from the β-decay of the pion which agrees with the prediction of the CVC theory. They calculate the diagram from the following interaction Hamiltonian

$$H_{\text{int}} = ig\left(\overline{\psi}_N \gamma_5 \tau_i \phi_i \psi_N\right) + 2^{-1/2} G \left(\overline{\psi}_\nu \gamma_\sigma (1 + \gamma_5) \psi_\mu\right) \left(\varepsilon_{1jk} - i\varepsilon_{2jk}\right) \cdot \left(\phi_j \partial_\sigma \phi_k\right)$$

where g is the renormalized pion-nucleon coupling constant and G, the weak vector-coupling constant. In the calculation of the matrix elements the static limit is used for nucleons which are then treated as a Fermi gas, and closure approximation is used in calculating the capture probability.

It is then assumed that a proton produced anywhere in the nucleus always arrives at the nuclear surface and the probability of its being transmitted through the nuclear surface is estimated by using transmission functions [62]. One obtains a proton emission rate of 1.7% for Ca^{40}, 1.2% for Ag^{107}, and 1.1% for ^{208}Pb. The energy spectrum calculated for capture in ^{107}Ag shows a peak around 25 MeV and extends up to proton energies of 80 MeV.

In the calculation of Bertero et al. [54] a particular diagram was singled out. However, to the same order in the interactions considered there are additional diagrams contributing to the process, which could interfere destructively with the one taken into account [63]. In fact, one knows that as a consequence of Siegert's theorem [64, 65] on the interaction between nucleons and the radiation field, there are no exchange terms contributing to the electric multipole transitions, and in the limit of zero photon energy they completely disappear due to the vanishing of the magnetic terms. Experimentally this result seems to hold also for fairly high ($\lesssim 100$ MeV) photon energies. On the basis of CVC theory, one expects an extended Siegert theorem to hold also for the vector part of the weak interaction [66]. Hence, the exchange contribution might be much smaller than calculated by Bertero et al. [54]. This has been recently verified experimentally by Kotelchuk and Tyler [51] as discussed in the following section.

3.4 Recent Experiments

In the last few years several new experiments on the detection of charged particles following muon capture have been performed.

Kotelchuk [46] analyzed an emulsion experiment [50] containing nearly 10^5 muon endings. He found evidence for muon capture in the nuclear surface [16] from the number of one-prong stars found, their expected number for surface or volume capture being deduced from neutron reactions at comparable energies. In a subsequent analysis [51] of the same plates, Kotelchuk and Tyler present convincing evidence against the preponderance of the mechanism of capture by exchange currents suggested by Bertero et al. [54]. The proton-energy spectrum, obtained from 1289 single-prong μ-stars, peaks at 7–8 MeV, as opposed to the peaking around 25 MeV predicted by Bertero et al. (Fig. 20). Also, much fewer protons than predicted are observed in the 20–40 MeV region, and there is complete absence of stars beyond 40 MeV, while more than 100 were to be found from the mechanism of Bertero et al. [54]. Kotelchuk and Tyler then conclude that this mechanism occurs less often than predicted at least by a factor $(2.0 \pm 1.0) \times 10^{-2}$. On the other side, the experimental spectrum is quite consistent with Singer's pseudodeuteron model [16].

Vaisenberg et al. [47], from an emulsion experiment, report 3% probability for charged particle emission following μ-capture, in agreement with previous results. In a high-statistics emulsion experiment involving $(2.74 \pm 0.14) \times 10^5$ stopped muons, Vaisenberg et al. [48] investigate the spectrum of protons with energies higher than 25 MeV (they have 87 events in this region). Their results agree very well, within the statistical accuracy, with Kotelchuk and Tyler on the fraction of protons with energies higher than 25 MeV emitted per muon capture, which is found to be $3.16 \pm 0.34 \times 10^{-4}$. However,

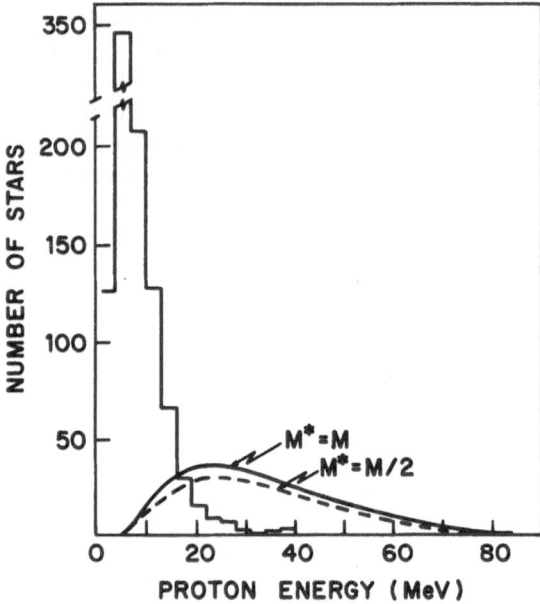

Fig. 20. Comparison of the experimental [51] proton energy spectrum with the predictions of Bertero et al. [54] (solid and dotted curves). The histogram represents the data corrected for α-particle emission and stars off light nuclei

they also find 13 events with protons having energies which extend up to ~80 MeV, amounting to $(4.7 \pm 1.1) \times 10^{-5}$ protons per capture, with energies beyond 40 MeV. They estimate that the possible admixture of pion stars in this energy region does not exceed 30%; hence this cannot explain away these high energy events. If it is then assumed that all the protons detected as having energies beyond 25 MeV are due to the Bertero mechanism [54], the latter will then account for only ~2% of the observed total proton emission. This agrees with the previous observation of Kotelchuk and Tyler [51] on the very small frequency of this mechanism.

So far we have discussed proton emission from the heavy nuclei of the nuclear emulsion. From these experiments there are also results on the percentage of charged particle emission per mu-capture from the light nuclei of the emulsion, namely C, N, and O. Morinaga and Fry have found [45] that, of muon capture in these light nuclei, 9.5% results in proton emission, and 3.4% in alpha emission. Vaisenberg et al. [47] confirm that the probability of emission of at least one charged particle after μ-capture in the light emulsion nuclei is ~15%.

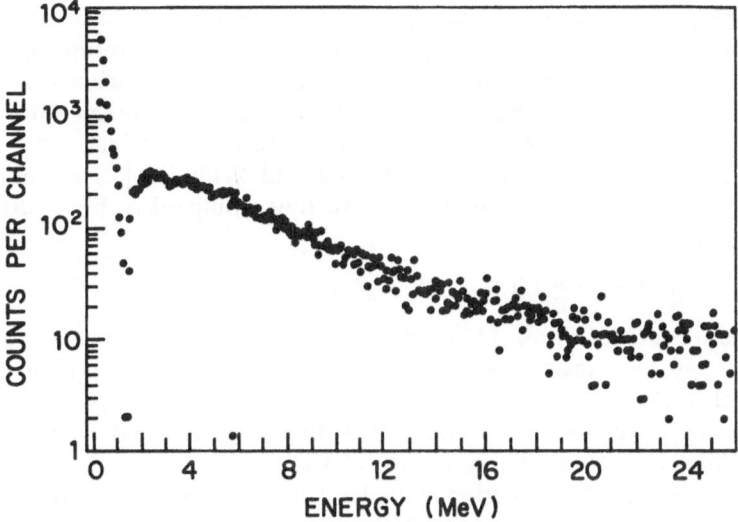

Fig. 21. The corrected energy spectrum for charged particle (supposedly mainly protons) emission following mu-capture in Si^{28} (Ref. [67])

Komarov and Savchenko [93], in a liquid-hydrogen with additions of neon bubble-chamber experiment, detect the charged particle emission following muon capture in another light nucleus, Neon, to be as high as $20 \pm 4\%$.

Sobottka and Wills [67] have performed a pioneering high-resolution experiment, in which the charged particles emitted after mu-capture in Si^{28} are detected by a Si(Li) scintillator. The spectrum of the charged particle emission, corrected for electron background and escape of protons, is given in Fig. 21. The low-energy end of the spectrum is identified as due to the Al^{27} recoiling nuclei from the $\mu^- + Si^{28} \rightarrow Al^{27} + n + \nu_\mu$ reaction, while the rest should be mainly due to the protons from $\mu^- + Si^{28} \rightarrow Mg^{27} + p + \nu_\mu$. There is however, no separation from other possible charged particles like α and d, which can be emitted by the excited Al^{28} nucleus. It should also be kept in mind that the protons can also come from $\mu^- + Si^{28} \rightarrow Mg^{26} + p + n + \nu_\mu$. The spectrum attributed by the authors mainly to protons (Fig. 21) has a low-energy cutoff at 1.4 MeV and a maximum at about 2.5 MeV, from which it decreases approximately exponentially with a decay constant of 4.6 MeV. The spectral integral gives 15% charged particle emission per mu-capture in Si^{28}. The figure is slightly higher than that obtained in emulsion experiments [45, 47] from the light C, O, N nuclei, but compares favourably with Komarov's [93] figure on Neon.

Vil'gel'mova et al. [68] have used an activation method to measure the probabilities of single proton emission after mu-capture in Si^{28} and K^{39}. In their experiment the final nucleus is identified from its radioactivity, their method thus providing a direct measurement for the Si$^{28}(\mu^-, p\nu_\mu)$Mg^{27} and K$^{39}(\mu^-, p\nu_\mu)$Cl^{38} reactions, which was not possible in the experiment of Sobottka and Wills [67]. They obtain for the percentage of single proton (unaccompanied by neutron) emission

$$W_{\mu^-, p\nu_\mu}(\text{Si}^{28}) = 5.3 \pm 1.0\%, \qquad W_{\mu^-, p\nu_\mu}(\text{K}^{39}) = 3.2 \pm 0.6\%. \tag{20}$$

If the results of Refs. [67] and [68] are sufficiently accurate, an interesting conclusion is that the reaction Si$^{28}(\mu^-, pn\nu_\mu)$Mg^{26} is probably as frequent as Si$^{28}(\mu^-, p\nu_\mu)$Mg^{27} and moreover, the 15% charged particle emission measured by Sobottka and Wills [67] probably contains a fair percentage of α and d.

As mentioned in Section 2.2, the Russian group had previously measured [24] the Si$^{28}(\mu^-, \nu_\mu)$Al^{28} reaction, for which they found a $28 \pm 4\%$ probability. Thus, the combined results for (μ^-, ν_μ) and $(\mu^-, p\nu_\mu)$ of the Russian group give $33 \pm 4\%$ probability for muon capture in Si^{28} without subsequent emission of neutrons. This agrees very well with the figure of $36 \pm 6\%$ for no-neutron emission obtained by MacDonald et al. [1, 69].

Qualitative confirmation of these results on proton emission, following μ-capture in Si^{28} is presented by Miller et al. [86] and by Temple et al. [85], who used the delayed $-\gamma$ method to identify the nature of the emitted particles. Temple et al. [85] have also evidence for charged particles emission after capture in Ca.

Budyashov et al. [70] have recently performed a counter experiment to study the charged particles following muon capture in several light and intermediate nuclei, with the specific aims of separating the charged particles by masses and measuring the energy spectra of the emitted protons, deuterons and tritons. The targets used in the experiment were ^{28}Si, ^{32}S, ^{40}Ca, and ^{64}Cu. The proton energy spectra are measured from $E_p = 15$ MeV and extend to ~ 60 MeV and those for deuterons are measured from $E_d = 18$ MeV and extend to ~ 50 MeV (Fig. 22). The yield of proton emission with energies above 15 MeV was found to be $0.88 \pm 0.06\%$ for ^{28}Si, $1.30 \pm 0.11\%$ for ^{40}Ca and $0.60 \pm 0.07\%$ for ^{64}Cu. The yield of deuteron emission with energies above 18 MeV obtains as $0.33 \pm 0.03\%$ for ^{28}Si, $0.22 \pm 0.03\%$ for ^{40}Ca and $0.10 \pm 0.03\%$ for ^{64}Cu. Their detailed results are presented in Figs. 23, 24. The amount of tritium was found to be much smaller. The decrease in the percentage of deuteron emission from the total yield of charged particles with the increase of the charge number

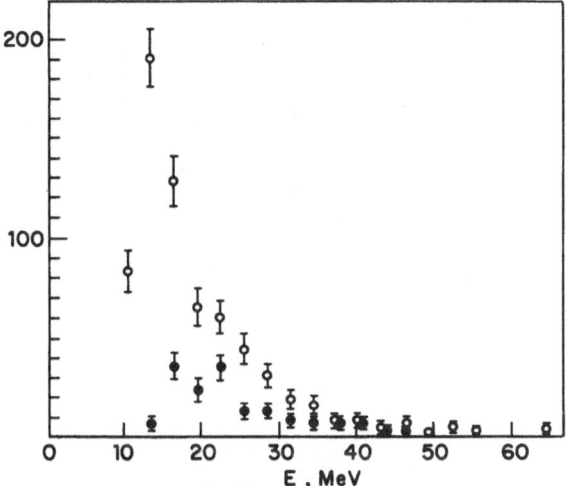

Fig. 22. The energy spectra of protons (O) and deuterons (●) from mu-capture in
^{32}S (Ref. [70])

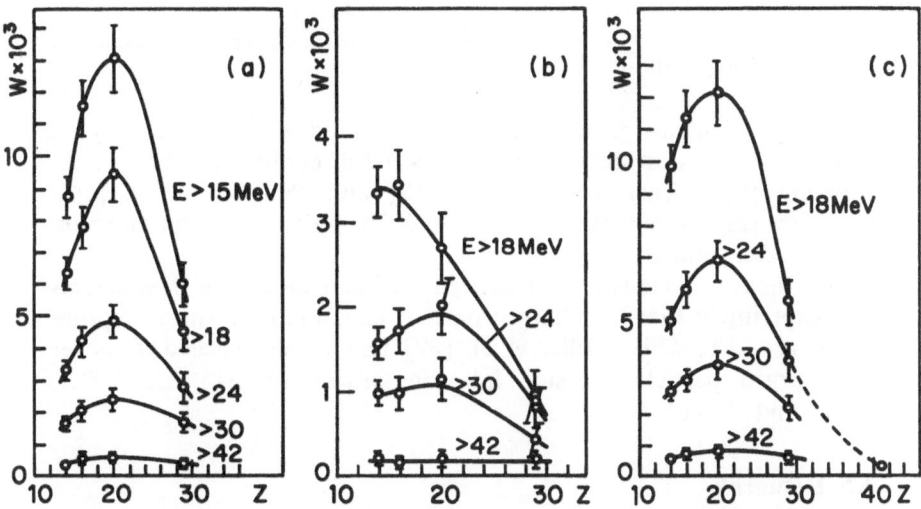

Fig. 23. Dependence of the integral probability of charged particle emission on the nuclear
charge, from the experiment of Ref. [70] for (a) protons, (b) deuterons, (c) total charged
particles

agrees with an earlier estimate from emulsion experiment [47]. The
ratio of deuteron to proton emission found in this experiment should
be of great value to the theoretical formulation of charged particle
emission in these nuclei.

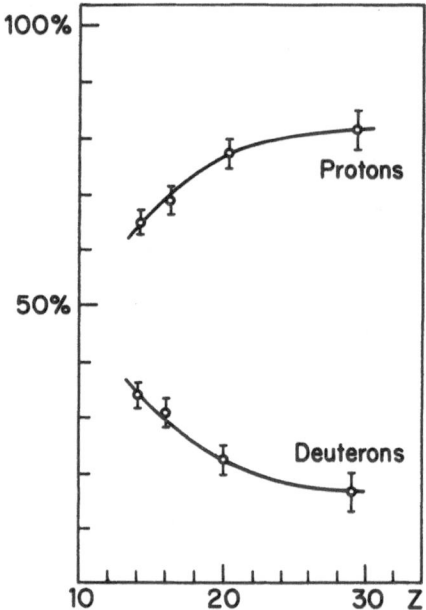

Fig. 24. Proton and deuteron yields as fractions of the total charged particle emission, for several nuclei (Ref. [70])

Before concluding, we should like to add that the scarcity of proton emission in heavy nuclei has been confirmed also in the experiments of Backenstoss et al. [18] and Petitjean et al. [19], who find respectively less than 1% and 3% possible proton emission in the nuclei studied.

Furthermore, Heusser and Kirsten [90] conclude from their activation experiment that the charged particle emission following μ-capture in Ni is $\sim 4\%$, while Miller et al. [87] find no evidence whatsoever for charged particle emission following μ-capture in ^{142}Ce, ^{140}Ce, ^{138}Ba, and ^{120}Sn.

3.5 Remarks

1. There is hardly any need to emphasize the lack of detailed theoretical attempts to treat the problem of charged particle emission from nuclei following mu-capture. With the experimental information which has been accumulating in the last few years, there is more ground now for testing possible theoretical approaches.

2. The existing experimental evidence [48, 51] appears to rule out the exchange current mechanism [54, 63] as the main contribution to the rate of proton emission from heavy nuclei.

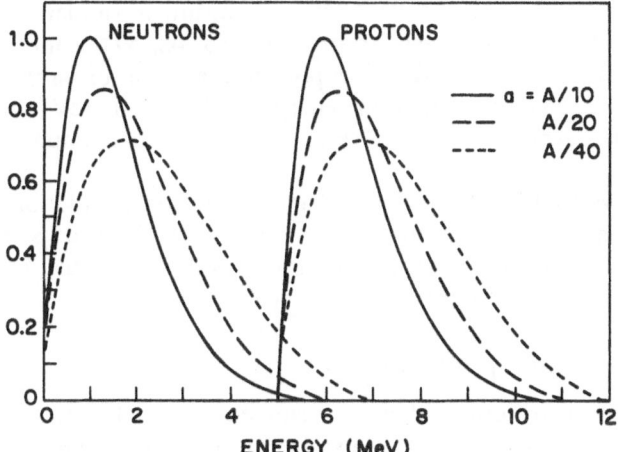

Fig. 25. Calculated evaporation spectra of neutrons and protons from Ge[68] at 20 MeV excitation energies (Refs. [53, 71])

3. The rate and the energy spectrum of the protons emitted after mu-capture in Ag, Br are in qualitative agreement [46–48, 51] with the pseudodeuteron model [16]. There is however no detailed prediction available of this model concerning the expected energy spectrum. Nevertheless, it should be remarked that the high energy protons (up to 80 MeV) observed by Vaisenberg et al. [48] are not unaccounted for in the pseudodeuteron model. The maximal energy of 50 MeV protons expected from the model is only a first approximation, as the motion of the pseudodeuterons in the nuclear surface would certainly smear out the 50 MeV limit.

4. It would be of great interest to measure the asymmetry of the emitted protons with respect to the muon polarization. Such an asymmetry would single out the pseudodeuteron model (or some other yet unthought of form of direct emission) versus the symmetric distribution expected for evaporation protons.

5. The calculation of Ishii [49] excludes the evaporation mechanism as the main producer of protons following mu-capture in Ag, Br. The observed spectra of the protons [45, 48, 51, 70] are also indicative of a direct mechanism, particularly in view of the amount of relatively high energy protons observed.

For comparison one can look at nuclear reactions at the appropriate energies. In Fig. 25 we show the calculated [53, 71] evaporation spectra of neutrons and protons from Ge[68] at 20 MeV excitation energies, the average excitation energy in intermediate and heavy nuclei after mu-capture being also 15–20 MeV. The neutron evaporation spectrum observed [5, 9] after mu-capture is similar to that in Fig. 25 while that observed for protons [45, 48, 52] definitely shifts towards higher energies. The evaporation spectra of protons from nuclear reactions at excitation energies of 15–20 MeV do follow [53, 71] the general shape of Fig. 25, with a maximum between 3–5 MeV.

It is nevertheless of interest to have additional calculations on the contribution of evaporation to the spectra of charged particles in various nuclei, as its contribution might still be sizable in certain nuclei, at least for the lower energy part of the spectrum.

IV. Neutron and γ Asymmetry

The angular distribution of direct neutrons emitted following muon capture in nuclei is expected to exhibit an asymmetry with respect to the direction of polarization of the captured muons. The asymmetry, related to parity violation in weak interactions carries information on the weak interaction coupling constants. Likewise, the γ-rays from radiative muon capture exhibit an angular asymmetry with respect to the μ-polarization. As these topics have also been discussed at this Conference by Prof. Überall and Dr. Rosenstein, we shall record here only briefly some recent developments on these topics.

The angular asymmetry of the neutrons can be expressed [72]

$$F(\theta) = 1 + P_\mu \alpha(E_n)\, \theta_{\mu n} . \tag{21}$$

P_μ is the degree of muon polarization, and $\alpha(E_n)$ contains both nuclear and weak-interaction information. In simple models one can separate $\alpha = \alpha_C \cdot \alpha_N$, where α_N contains the nuclear matrix elements, and α_C contains only the weak interaction coupling constants. Primakoff's [73] formula for α_C in spin-0 nuclei gives

$$\alpha_C = \frac{G_V^2 - G_A^2 - 2G_A G_P + G_P^2}{G_V^2 + 3G_A^2 + G_P^2 < 2G_A G_P} \simeq -0.4 . \tag{22}$$

The inclusion of momentum dependent terms has been shown [39] to increase α_C to -0.11. α_N depends fairly strongly on E_n and its calculation, resulting in values between $0.2-0.8$, is quite model-dependent [72].

Early measurements [74] of α in Ca and Si indicated negative values close to -1. A more recent experiment by Sculli [75] shows the possibility of positive α in Ca (Fig. 26). Sundelin et al. [76, 33] have recently measured the asymmetry parameter for capture in Si, S, and Ca, for neutron energies ranging between 10 and 45 MeV. They obtain for all nuclei studied positive values for α which generally increase with neutron energy (Figs. 27–31).

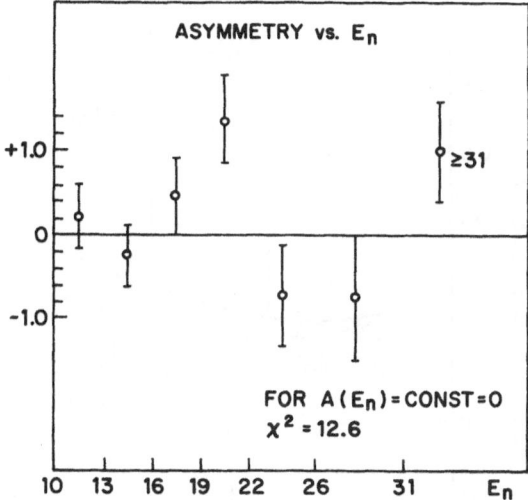

Fig. 26. The neutron asymmetry as a function of neutron energy from mu-capture in Ca^{40} (Ref. [75])

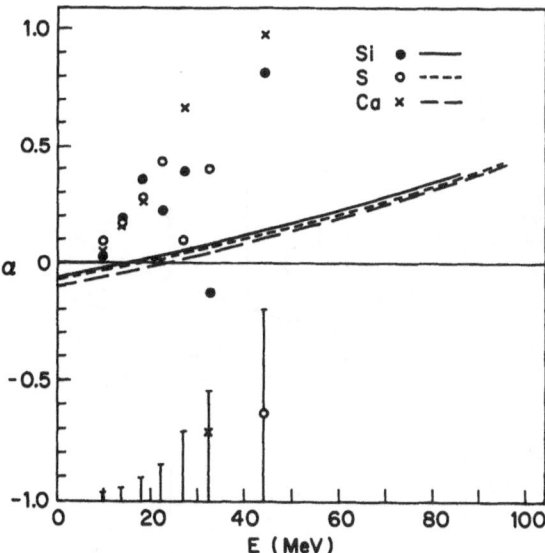

Fig. 27. Calculation of Bogan [37] of the neutron asymmetry coefficient compared to the measurements of Sundelin et al. [76]. The standard deviation of the experimental data is shown at the bottom of the figure

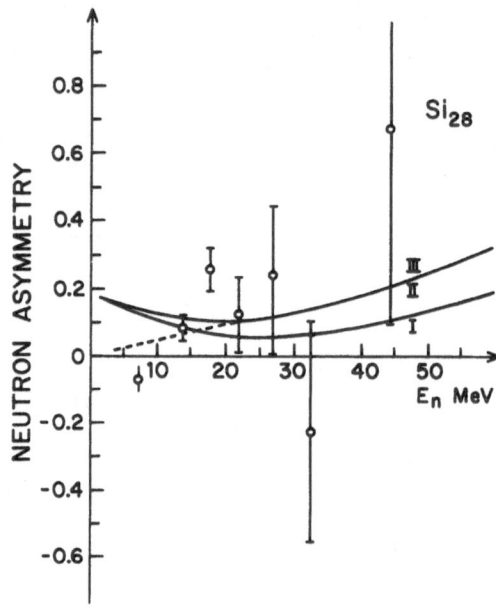

Fig. 28. Calculations [38] of the neutron asymmetry coefficient in Si^{28} compared to Sundelin's measurement [76]

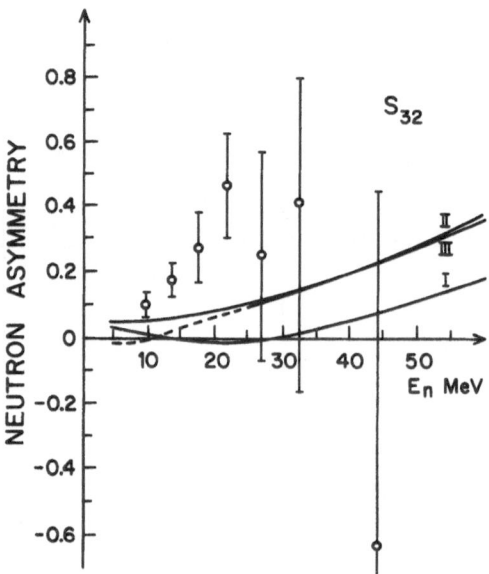

Fig. 29. Calculations [38] of the neutron asymmetry coefficient in S^{32} compared to Sundelin's measurement [76]

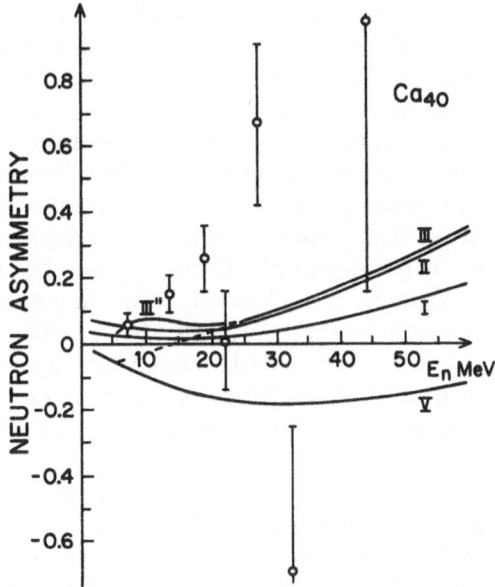

Fig. 30. Calculations [38] of the neutron asymmetry parameter in Ca^{40} compared to Sundelin's measurement [76]

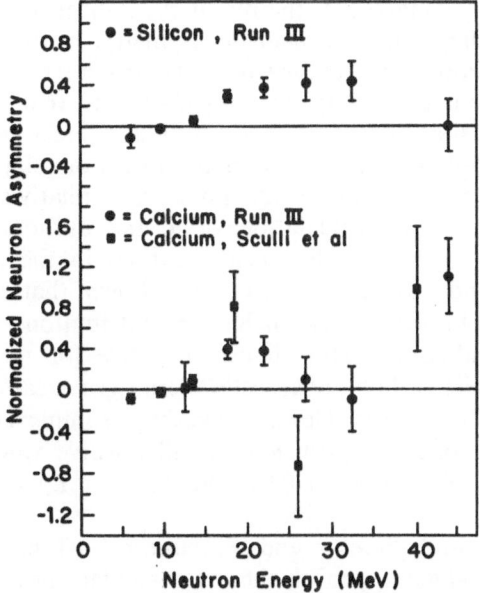

Fig. 31. The neutron asymmetry parameter for Si and Ca from the second experiment of Edelstein and Sundelin (Ref. [33])

In Figs. 27–30 some theoretical calculations are compared with the older data [76] of Edelstein and Sundelin, while in Fig. 31 their new data (the second of Ref. [33]) are presented. There is general agreement between the two sets of data, the newer experiment [33] enjoying, however, much better statistics. The later experiment also gives additional evidence to the apparent energy variation of α, which seems to reach maximal positive values of ~ 0.4 for both Si and Ca around 20–30 MeV, then possibly decreasing for higher energies.

Also, in a new experiment [94], Evseev's group has remeasured the polarization coefficients in Calcium and Sulphure. Their new values are $\alpha_{Ca} = 0.035 \pm 0.11$ (average over the energy range $2 \text{ MeV} < E_n < 17 \text{ MeV}$) and $\alpha_S = 0.015 \pm 0.025$ (average over the energy range $2.5 \text{ MeV} < E_n < 10 \text{ MeV}$). These values are not in disagreement any more with the results of Edelstein and Sundelin [33, 76] in these energy ranges.

Positive values for α were unexpected from earlier estimates [39]. However, recent calculations by Bogan [37] and Piketty and Procureur [38], who include relativistic effects up to order $1/M$ [37] and $1/M^2$ [38] in their shell model treatment of the process, show that positive α's are thus obtainable [82]. Their results (Figs. 27–30), although in the right direction, are generally smaller than Sundelin's [33, 76] values.

A comparative analysis of the recent theoretical papers [37–39, 80, 82, 95–97] dealing with the neutron asymmetry, leads to certain conclusions on the importance of the various factors entering the calculation of α. Thus, there is general agreement [37–39, 96] that the inclusion of relativistic effects, as suggested firstly by Klein, Neal and Wolfenstein [39] will increase the value of α in the direction of positive values. Keeping terms up to order P_n/M (first-order relativistic correction) in a Fermi-gas model calculation of α, the authors of Ref. [39] obtained values which are less negative than in the non-relativistic case. Piketty and Procureur [38] have shown that additional improvement is obtained by also including contributions to order M^{-2}, although the values are still slightly negative for neutron energies $E_n > 5 \text{ MeV}$, while a fully relativistic Fermi-gas treatment does not bring any further changes. Nevertheless, by assuming effective masses $M_p^* = M_n^* = M/2$ one can push α to small positive values. For ^{40}Ca, α then changes [96] between 0.05 and 0.02, for E_n varying between 0 and 30 MeV.

Bogan [37] and Piketty and Procureur [38] have shown that positive values, which approach the experimental ones, are obtainable by using a realistic nuclear model. In both these papers shell-model wave functions are used with the harmonic oscillator parameters fitted to give the proper r.m.s. radius [37] or from $p - 2p$ and $e - ep$ experiments [38]. The use of the more suitable nuclear model combined with the inclusion of relativistic terms leads to the positive values given in Figs. 27–30, and it should also be mentioned that the results are not sensitive to reasonable changes in the parameters of the nuclear model. On the other hand, increasing the weak pseudoscalar coupling (whose value is not very well established) by 20%, increases [38] the asymmetry by some 30%. In these papers the final state interaction is taken into account as described in Section 2.3, but the authors concluded that it has little effect on the neutron asymmetry. A similar conclusion is reached [38] concerning the spin-orbit interaction, although on this point there is no confirmation from the work of Bouyssy et al. [97], whose calculation is moderately sensitive to such a term. The model of Madurga [92] gives for α in Ca a monotonically increasing function of energy, changing between 0.09 and 0.23 for E_n between 10 and 50 MeV.

Bouyssy and Vinh Nau have also checked [82] the insensitivity of the asymmetry of neutrons emitted after polarized muon capture in O^{16} and Ca^{40} to various nuclear models, by considering a pure shell-model and multi-hole-multi-particle configuration-mixing nuclear wave functions. On the other hand, in contradiction to previous conclusions [37, 38], they find in their treatment of the final state interaction that the addition of a surface term to the imaginary part of the optical potential affects considerably the energy dependence of α. In fact, they find an energy dependence (Fig. 32, curve 6) which reproduces fairly well the magnitude and overall behaviour of $\alpha(E_n)$, as given by Edelstein and Sundelin [33] (Fig. 31). However, the neutron intensity spectrum obtained by using the parameters used in calculating curve 6 of Fig. 32 compares poorly with the experimental findings. The agreement is markedly improved when they use [91] for Ca the wave functions of Campi and Spring [94], which are Hartree-Fock wave functions for a density dependent effective nucleon-nucleon interaction, while at the same time the asymmetry parameter does not change significantly from that given by curve 6 of Fig. 32.

Eramzhyan and Salganic [99] confirm large effects of the final state interaction on the neutron asymmetry in a calculation of $\mu + {}^{16}O \rightarrow {}^{15}N + n + \nu$. They consider the transitions to the ground state of ${}^{15}N$ and to three of its excited levels. Using the distorted wave approximation for the final state, they find marked differences compared to a plane wave approximation, the asymmetry becoming an oscillating function of E_n. This feature is quite stable against slight changes in the parameters of the optical potential. Thus, the question

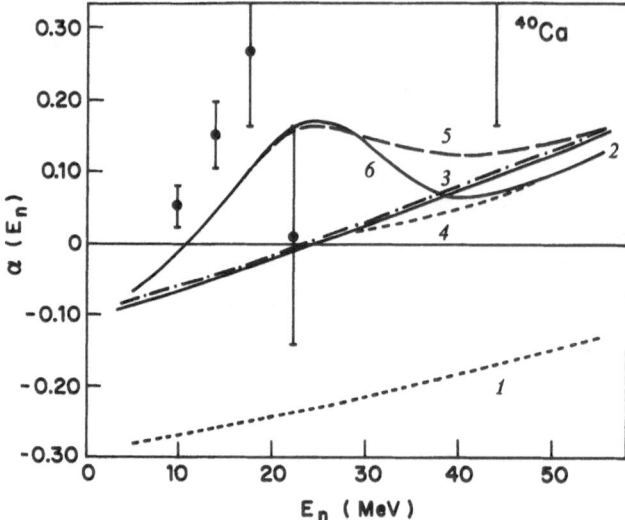

Fig. 32. Calculated (Ref. [82]) neutron asymmetry coefficient for Ca^{40}. Curve *1*: shell-model without relativistic corrections and without f.s.i. Curve *2*: same model with first order relativistic terms and real optical potential. Curve *3*: same as curve *2* but in configuration mixing model. Curve *4*: same as curve *2* plus volume absorption. Curve *5*: same as curve *2* plus surface absorption. Curve *6*: same as curve *2* plus volume and surface absorption

of the sensitivity of α to the final state interaction is certainly an open one so far.

Turning now to the γ-asymmetry, we recall [77] that γ-rays from the radiative capture $\mu^- + p \rightarrow n + \nu_\mu + \gamma$ should be emitted because of parity violation with an asymmetric distribution

$$N(\theta) = 1 + P_\mu \alpha_\gamma \cos\theta_{\gamma\mu} , \qquad (23)$$

where P_μ is the muon polarization, $\theta_{\gamma\mu}$ the angle between the γ-momentum and the μ spin polarization. The asymmetry parameter α_γ is equal to the γ-ray circular polarization, if only the muon radiation is taken into account and for $V-A$ interaction one has $\alpha_\gamma = 1$. Nuclear effects reduce α_γ in radiative muon capture in nuclei, and Rood and Tolhoek [78] have calculated $\alpha_\gamma = 0.75$ for radiative muon capture in Ca^{40} averaged over the γ-energy interval between 57–75 MeV.

In a recent experiment [79] discussed at this conference by Dr. Rosenstein, Di Lella et al. report a measurement of the γ-asymmetry for γ-energies between 57–75 MeV as giving

$$\alpha_\gamma \leqq -0.32 \pm 0.48 . \qquad (24)$$

The equality holds for vanishing neutron asymmetry parameter $\alpha_n = 0$, the knowledge of which is required in order to separate the neutron contamination. If Sundelin's [76] value for $\alpha(E_n \simeq 50 \text{ MeV})$ is used, then the value of α_y could be as low as -0.6 ± 0.5. This is in gross disagreement with theoretical expectations. A recent calculation of Longuemare and Piketty [80] of the neutron asymmetry parameter for the most energetic neutrons emitted after μ-capture finds for its limiting value $\alpha = 0.70$, quite independent of the model used. Their result thus aggravates the puzzle raised by the experiment of DiLella, Hammerman and Rosenstein.

References

1. MacDonald, B., Diaz, J. A., Kaplan, S. N., Pyle, R. V.: Phys. Rev. 139, B 1253 (1965). Earlier experimental work on the average neutron emission is referred to here in Table V.
2. Ball, W., Lauterjung, K. H.: Z. Naturforsch. 8 A, 214 (1953).
3. Sundelin, R. M., Edelstein, R. M., Suzuki, A., Takahashi, K.: Phys. Rev. Letters 20, 1198 (1968).
4. Krieger, M. H.: Thesis, Columbia University, NEVIS-172 (1969).
5. Schröder, W. U.: Thesis, Technische Hochschule Darmstadt (1971); – Jahnke, U. et al.: CERN preprint; – Schröder, W. U., Jahnke, U., Lindenberger, K. H., Röschert, G.: Darmstadt preprint (May 1973).
6. Balashov, V. V., Belyaev, V. B., Kabachnik, N. M., Eramzhian, R. A.: Phys. Letters 9, 168 (1964).
7. Foldy, L. L., Walecka, J. D.: Nuovo Cimento 34, 1026 (1964).
8. Überall, H.: Suppl. Nuovo Cimento 4, 781 (1966).
9. Evseev, V., Kozlovski, T., Roganov, V., Woitkowska, J.: In: Devons, S. (Ed.): High-energy physics and nuclear structure, p. 157. New York: Plenum Press 1970. – Wojtkowska, J., Evseev, V. S., Kozlovski, T., Mamedov, T. N., Roganov, V. S.: Yad. Fiz. 14, 624 (1971) [Soviet. J. Nucl. Phys. 14, 349 (1972)]. – Wojtkowska, J., Evseev, V. S., Kozlovski, T., Roganov, V. S.: Yad. Fiz. 15, 1154 (1972) [Sov. J. Nucl. Phys. 15, 639 (1972)].
10. Tiomno, J., Wheeler, J. A.: Rev. Mod. Phys. 21, 153 (1949).
11. Kaplan, S. N., Moyer, B. J., Pyle, R. V.: Phys. Rev. 112, 968 (1958).
12. Brueckner, K. A., Gammel, J. L.: Phys. Rev. 109, 1023 (1958).
13. Singer, P.: Nuovo Cimento 23, 669 (1962).
14. Brueckner, K. A., Lockett, A. M., Rotenberg, M.: Phys. Rev. 121, 255 (1961).
15. Weisskopf, V. F., Ewing, D. H.: Phys. Rev. 57, 472 (1940).
16. Singer, P.: Phys. Rev. 124, 1602 (1961).
17. Turkevich, A., Fung, S. C.: Phys. Rev. 92, 521 (1953). – Winsberg, L.: Phys. Rev. 95, 205 (1954).
18. Backenstoss, G., Charalambus, S., Daniel, H., Hamilton, W. D., Lynen, U., v. d. Malsburg, Ch., Poelz, G., Povel, H. P.: Nucl. Phys. A 162, 541 (1971).
19. Petitjean, C., Backe, H., Engfer, R., Jahnke, U., Lindenberger, K. H., Schneuwly, H., Schröder, W. U., Walter, H. K.: Nucl. Phys. A 178, 193 (1971).
20. Kessler, D., McKee, R. J., Hargrove, C. K., Hincks, E. P., Anderson, H. L.: Canad. J. Phys. 48, 3029 (1970).
21. Pratt, T. A. E. C.: Nuovo Cimento 61 B, 119 (1969).
22. Earle, E. D., Bartholomew, G. A.: Nucl. Phys. A 176, 363 (1971).
23. Bunatyan, G. G., Evseev, V. S., Nikityuk, L. N., Nikolina, A. A., Pokrovskii, V. N., Roganov, V. S., Smirnova, L. M., Yutlandov, I. A.: Yad. Fiz. 9, 783 (1969); [Soviet J. Nucl. Phys. 9, 457 (1969)].

24. Bunatyan,G.G., Evseev,V.S., Nikityuk,L.N., Pokrovskii,V.N., Rybakov,V.N., Yutlandov,I.A.: Yad. Fiz. **11**, 795 (1970); [(Soviet J. Nucl. Phys. **11**, 444 (1970)].
25. Raphael,R., Überall,H., Werntz,C.: Phys. Letters **24**B, 15 (1967).
26. Kelly,F.J., Überall,H.: Nucl. Phys. A**118**, 302 (1968).
27. See also Pratt,T.A.E.C.: Nuovo Cimento **68**A, 477 (1970).
28. Evseev,V., Kozlowski,T., Roganov,V., Wojtkowska,J.: Phys. Letters **28**B, 553 (1969).
29. Plett,M.E., Sobottka,S.E.: Phys. Rev. C**3**, 1003 (1971).
30. Singer,P., Zin,A.: to be published.
31. Hagge,D.: Thesis. University of California, UCRL–10516 (1963).
32. Turner,L.: Thesis. Carnegie Institute of Technology, CAR-882-5 (1964).
33. Sundelin,R.M.: Thesis. Carnegie Institute of Technology, CAR-882-22 (1967). – Sundelin,R.M., Edelstein,R.M.: Phys. Rev. C**7**, 1037 (1973).
34. Überall,H.: Nuovo Cimento **6**, 533 (1957).
35. Lubkin,E.: Ann. Phys. **11**, 414 (1960).
36. Dolinsky,E.I., Blokhintsev,L.D.: Nucl. Phys. **10**, 527 (1959).
37. Bogan,A.: Phys. Rev. Letters **22**, 71 (1969). – Nucl. Phys. B**12**, 89 (1969).
38. Piketty,C.A., Procureur,J.: Nucl. Phys. B**26**, 390 (1971).
39. Klein,R., Neal,T., Wolfenstein,L.: Phys. Rev. **138**, B 86 (1965).
40. Akimova,M.K., Blokhintsev,L.D., Dolinskii,E.I.: J. Exptl. Theoret. Phys. (U.S.S.R.) **39**, 1806 (1960); [Sov. Phys. JETP **12**, 1260 (1961)].
41. Novikov,V.M., Urin,M.G.: Yad. Fiz. **6**, 1233 (1967); [Soviet J. Nucl. Phys. **6**, 898 (1968)].
42. Fujii,A., Primakoff,H.: Nuovo Cimento **12**, 327 (1959).
43. Moniz,E.J.: Phys. Rev. **184**, 1154 (1969). – Moniz,E.J., Sick,I., Whitney,R.R., Ficenec,J.R., Kephart,R.D., Trower,W.P.: Phys. Rev. Letters **26**, 445 (1971).
44. Gilbert,A., Cameron,A.G.W.: Canad. J. Phys. **43**, 1446 (1965).
45. Morinaga,H., Fry,W.F.: Nuovo Cimento **10**, 308 (1953).
46. Kotelchuck,D.: Nuovo Cimento **35**, 27 (1964).
47. Vaisenberg,A.O., Kolganova,E.D., Rabin,N.V.: Yad. Fiz. **1**, 652 (1965) [Soviet J. Nucl. Phys. **1**, 467 (1965)].
48. Vaisenberg,A.O., Kolganova,E.D., Rabin,N.V.: Yad. Fiz. **11**, 830 (1970) [Soviet J. Nucl. Phys. **11**, 464 (1970)].
49. Ishii,C.: Prog. Theor. Phys. **21**, 663 (1959).
50. Kotelchuck,D., McEwen,J.G., Orear,J.: Phys. Rev. **129**, 876 (1963).
51. Kotelchuck,D., Tyler,J.V.: Phys. Rev. **165**, 1190 (1968).
52. Endt,P.M., Demeur,M., (Ed.): Nuclear reactions, Vol. I, Chaps. VII, IX. Amsterdam: North Holland Publ. Co. 1959.
53. Kawai,M., Kikuchi,K.: Nuclear matter and nuclear reactions. Amsterdam: North Holland Publ. Co. 1968.
54. Bertero,M., Passatore,G., Viano,G.A.: Nuovo Cimento **38**, 1669 (1965).
55. Wilkinson,D.H., In: Birks,J.B., (Ed.): Proc. of the Rutherford Jubilee Inter. Conf., p. 339 London: Heywood & Co. 1961.
56. Tagami,T.: Prog. Theor. Phys. **21**, 465 (1959).
57. da Providencia,J.: Proc. Phys. Soc. **77**, 81 (1961).
58. Überall,H., Wolfenstein,L.: Nuovo Cimento **10**, 136 (1958).
59. Hodgson,P.E.: Nucl. Phys. **8**, 1 (1958).
60. Elton,L.R.B., Gomes,L.C.: Phys. Rev. **105**, 1027 (1957).
61. Oda,N., Harada,K.: Nucl. Phys. **7**, 251 (1958).
62. Feshbach,H., Shapiro,M.M., Weisskopf,V.F.: USAEC Report NY03077 (1953).
63. Bertero,M., Passatore,G., Viano,G.A.: Nuovo Cimento **52**, 1379 (1967).
64. Siegert,A.J.F.: Phys. Rev. **52**, 787 (1937).
65. Sachs,R.G., Austern,N.: Phys. Rev. **81**, 705 (1951).
66. See, e.g. Schopper,H.F.: Weak interactions and nuclear beta decay. Amsterdam: North Holland Publ. Co., 1966.

67. Sobottka, S. E., Wills, E. L.: Phys. Rev. Letters **20**, 596 (1968).
68. Vil'gel'mova, L., Evseev, V. S., Nikityuk, L. N., Pokovskii, V. N., Yutlandov, I. A.: Yad. Fiz. **13**, 551 (1971); [Soviet J. Nucl. Phys. **13**, 310 (1971)].
69. See also Charalambus, S.: Nucl. Phys. A **166**, 145 (1971).
70. Budyashov, Yu. G., Zinov, V. G., Konin, A. D., Mukhin, A. I., Chatrchyan, A. M.: Zh. Eksp. Teor. Fiz. **60**, 19 (1971) [Soviet Physics JETP **33**, 11 (1971)].
71. Dostrovsky, I., Fraenkel, Z., Weinberg, L.: Phys. Rev. **118**, 781 (1960).
72. See, e.g. Rho, M.: Lectures at Les Houches Summer School of Theoretical Physics (1968).
73. Primakoff, H.: Rev. Mod. Phys. **31**, 802 (1959).
74. Evseev, V., Roganov, V., Chernogorova, V., Chang Jun-wa, Szymczak, M.: Phys. Letters **6**, 332 (1963). – Evseev, V., Kilbinger, F., Roganov, V., Chernogorova, V., Szymczak, H.: Yad. Fiz. **4**, 545 (1967) [Soviet J. Nucl. Phys. **4**, 387 (1967)].
75. Sculli, J.: Thesis. Columbia University (1969).
76. Sundelin, R. M., Edelstein, R. M., Suzuki, A., Takahashi, K.: Phys. Rev. Letters **20**, 1201 (1968).
77. A detailed account on radiative muon capture can be found in Rood, H. P. C.: Thesis. Rijksuniversiteit te Groningen (1964).
78. Rood, H. P. C., Tolhoek, H. A.: Nucl. Phys. **70**, 658 (1965); see also Borchi, E., De Gennaro, S.: Phys. Rev. C **2**, 1012 (1970), who obtain slightly larger values for α_y.
79. DiLella, L., Hammerman, I., Rosenstein, L. M.: Phys. Rev. Letters **27**, 830 (1971).
80. Longuemare, C., Piketty, C. A.: Phys. Letters **38** B, 125 (1972).
81. Überall, H.: Springer Tracts in Mod. Phys. (to be published).
82. Bouyssy, A., Vinh Mau, N.: Nucl. Phys. A **185**, 32 (1972).
83. Bobodyanov, I.: Acta Phys. Acad. Sci. Hung. **29**, Supple. 4, 151 (1970).
84. Evans, H. J.: Nucl. Phys. A **207**, 379 (1973).
85. Temple, L. E., Kaplan, S. N., Pyle, R. V., Valley, G. F.: LBL 24 (1971). – Temple, L. E.: Thesis, LBL 781 (1972).
86. Miller, G. H., Eckhause, M., Martin, P., Welsh, R. E.: Phys. Rev. C **6**, 487 (1972).
87. Lucas, G. R., Jr., Martin, P., Miller, G. H., Welsh, R. E., Jenkins, D. A., Powers, R. J., Kunselman, A. R.: Phys. Rev. C **7**, 1678 (1973).
88. Bunatyan, G. G., Vie'gel'mova, L., Evseev, V. S., Nikityuk, L. N., Pokrovskii, V. W., Rybakov, V. N., Yutlandov, I. A.: Yad. Fiz. **15**, 945 (1972) [Soviet J. Nucl. Phys. **15**, 526 (1972)].
89. Kaplan, S. N.: In: Devons, S., (Ed.): High energy physics and nuclear structure, p. 143. New York: Plenum Press.
90. Heusser, G., Kirsten, T.: Nucl. Phys. A **195**, 369 (1972).
91. Bouyssy, A., Ngo, H., Vinh Mau, N.: Phys. Letters **44** B, 139 (1973).
92. Madurga, G.: Nuovo Cimento **12** A, 451 (1972).
93. Komarov, V. I., Savchenko, O. V.: Yad. Fiz. **8**, 415 (1968) [Soviet J. Nucl. Phys. **8**, 239 (1969)].
94. Wojtkowska, I., Evseev, V. S., Kozlowski, T., Nikolina, A. A., Roganov, V. S.: Yad. Fiz. **15**, 939 (1972) [Soviet J. Nucl. Phys. **15**, 523 (1972)].
95. Madurga, G.: Nuovo Cimento **12** A, 451 (1972).
96. Galindo, A., Pascual, P., Pascual, R.: Anal. Fisica (Spain) **68**, 275 (1972).
97. Bouyssy, A., Ngo, H., Vinh Mau, N.: Phys. Letters **44** B, 139 (1973).
98. Campi, X., Sprung, D. W. L.: Nucl. Phys. A **194**, 401 (1972).
99. Eramzhyan, R. A., Salganic, Yu. A.: Nucl. Phys. A **207**, 609 (1973).

Prof. Dr. Paul Singer
Department of Physics
Technion – Israel Institute of Technology
Haifa, Israel

The Two and Three Body Problem

J. S. Levinger*

Contents

---

* *Dedication*: I dedicate this report to my dear wife Gloria, and son Jody, who graciously consented to leave East Greenbush, New York and live in Paris for a year, so that I could work on this manuscript at the Institut de Physique Nucléaire, Orsay.   J. S. Levinger

I. Two-Nucleon and Three-Nucleon Systems

1.1 Introduction

The decade of the 1950's witnessed rapid progress in the phenomenology of the two-nucleon system, due particularly to measurements of proton-proton triplet scattering parameters. The basic technique was established; but very difficult extensions to the neutron-proton system remained. Theory did not keep up. The one-pion-exchange potential (OPEP) was reintroduced for large nucleon-nucleon distance with experimental verification for corresponding states of high angular momentum. Many attempts to extend field theory calculations led to uncertain results; perhaps these attempts are just now achieving some success.

    The decade of the 1960's witnessed rapid progress both in experimental and in theoretical work on the three-nucleon problem (Watson, 1967; Amado, 1969; Mitra, 1969; Delves, 1969; Sitenko, 1971; Delves, 1972; Harms, 1972; Slaus et al., 1972). Experimentalists made a variety of measurements on nucleon-deuteron scattering, and also studied the ground states of the trinucleon (the nuclei ^{3}H and ^{3}He) by electron scattering. Meanwhile many different theorists developed a variety of calculational methods for evaluating the energy of the triton, E_T. These same methods give the trinucleon wave function, and therefore form factors to compare with electron scattering. Nucleon-deuteron scattering at low energy has been calculated by several groups; but we are just beginning to be able to calculate $n - d$ (elastic and inelastic) scattering at higher energies.

I use the term "decade" loosely: I feel free to include 1972–1973 results in the "decade of the 1960's". Clearly, the work of neither decade is complete. Measurements and phenomenological fits to nucleon-nucleon scattering continue; and so do calculations on the ground state of the trinucleon. I have the *impression* that most of this work involves "evaluating the next decimal point". So this seems like a good time to review the successful calculations of the past decade.

In this review, I will attempt to give moderately partisan critiques of various approaches. This "moderation" will fail at times; because of my lack of sufficient knowledge of work in this field, or because of my prejudices. I am deliberately emphasizing the approach using a separable approximation to the two-nucleon t-matrix not because it is necessarily the best method, but simply because I know it best. I will discuss its successes and its failures. I will only summarize other methods, and give a partial guide to their literature.

As nuclear physicists, we are interested in the two-nucleon and three-nucleon problems mainly as first important steps in more difficult calculations for heavier nuclei. I shall first spend a little time discussing the one-nucleon problem. We hope to learn how to count. If you think that nuclear matter is the next tractable problem, you follow Gamow's title and count "1, 2, 3, infinity". Alternatively, you might want to tackle finite spherical nuclei, and count "1, 2, 3, 4, 16, 40, ...". When we examine different methods for mass number 3, we will try to find a method that will let us count past 3 – the supposed upper limit for the maligned Australian bushmen. (I believe that they could count higher; but that the European settlers weren't able to learn the Australian words and therefore claimed that they didn't exist. I say this since I find myself trying to take similar shortcuts when I deal unsuccessfully with a foreign language!)

One main theme of my review is that nature has presented us with a mélange of problems involving systems of one, two and three nucleons, and associated pion and other fields. I frequently refer to this mélange as a "Gordian knot" which I do not know how to cut (Levinger, 1973). I merely pick away at a loose strand here or there. Let's look at some of the tangle.

1.2 "One-Nucleon Problem"

First, the one-nucleon problem. Why is the neutron heavier than the proton? How can we calculate the mass difference? Presumably we need to understand each nucleon as a composite system. We obtain some evidence on the structure of the proton by measurements of the form factors for elastic electron-proton scattering (Wilson, 1964; Gourdin, 1966; Iachello, 1973). We find below that we need these form factors to interpret experiments on electron-deuteron and electron-trinucleon scattering. How will we find the neutron form factors which we also need? (It is circular reasoning to determine neutron form factors in the same experiments in which we hope to learn deuteron and trinucleon wave functions!)

Since the nucleon has structure, we should not be surprised to find "polarization" effects when one nucleon is close to another. These polarization effects (distortion of a meson cloud for instance) lead to the suggestion that we should anticipate non-additive effects when two nucleons are close to each other; e.g., the magnetic moment of the deuteron would not be just the sum of neutron and proton magnetic moments (with a well understood extra term for the D-state admixture). Further, when two nucleons are close together and polarize each other, we should not be surprised that this polarization changes the force these nucleons exert on a third nucleon nearby. Thus we are led to the expectation that there will be some "three-body force" which cannot be determined from experiments on two-nucleon scattering.

Three-body forces are familiar in molecular physics. Thus the potential of a system of three hydrogen atoms is not just the sum of two body potentials: in this case the two-body force "saturates". We have to be careful of our notation. Non-relativistically the electron-electron, electron-proton and proton-proton Coulomb forces are all pure two-body forces. We could in principle calculate the properties of a system of three hydrogen atoms by solving a *six-body problem* with only *two-body forces*. But if we wish to simplify by considering the *three-atom problem*, then we have to "pay" for this by introducing *three-body forces*. Since the nucleon has structure the trinucleon may have some similarity to a system of three hydrogen atoms.

1.3 Nucleon-Nucleon Scattering

Scattering measurements on the two-nucleon system, give scattering amplitudes, which can be expressed either as phase shifts, or as on-shell elements of the transition matrix, called the t-matrix. (For simplicity of notation, we consider central forces at present for a specified partial wave.) By "on-shell" we mean that we are considering elastic scattering: the original magnitude of the relative linear momentum, p, equals the final magnitude k. Further, p and k are each related to the energy E in the standard way for non-relativistic kinetic energy: $E = p^2 = k^2$. (Throughout this review I use units in which $\hbar = M = 1$, where M is the nucleon mass.)

We can write matrix elements of the t-matrix between the real values of linear momentum before and after scattering. Thus we write $\langle p | t(E) | k \rangle$, or alternatively $t(p, k; E)$. If we know $\mathbf{t}(E)$ as a matrix (bold face roman shows a matrix) we could go from the diagonal elements where $p = k$ to off-diagonal elements, $p \neq k$. If we keep the energy equal to the square of either momentum, say $E = k^2 \neq p^2$, then we are dealing with a "half-off-shell" element of the t-matrix. On the other hand if we use the Lippmann-Schwinger (LS) equation to go from the potential v to the matrix $\mathbf{t}(E)$, we can also find the matrix for complex energy s. Matrix elements $t(p, k; s)$ for which we have the three inequalities given below are called "completely off-shell".

$$p \neq k, \quad \text{and}$$
$$s \neq k^2, \quad \text{and} \tag{1.1}$$
$$s \neq p^2.$$

(In the important case of energy s real and negative, the latter two equations are automatically satisfied.)

We will review, but not derive, the mathematics involved (e.g., the LS equation) in Chapter III. Here I'd like to give an *intuitive introduction* to our reasons for being interested in off-shell values of the t-matrix. We can go half-off-shell by nucleon-nucleon bremsstrahlung. The incident energy E and incident momentum k are, for a two-body system, related by $E = k^2$. But in the final state the photon has momentum, so we see that the final *nucleon* momentum $p \neq k$.

We find in Chapter V that we must go completely off-shell in the Faddeev approach to the three-nucleon system. Why? Consider two nucleons that scatter each other in the presence of a third "spectator" nucleon. The energy of the entire system is negative (for the trinucleon bound state) and the third nucleon has some positive energy: hence the energy of the two-nucleon system s is negative. Also, the two nucleons need not conserve momentum in this scattering; since the third nucleon can take up some momentum. Hence in general $p \neq k$, for the two-nucleon system. Thus we are concerned with elements of the t-matrix obeying 1.1.

We frequently use a separable approximation to the t-matrix. (See Chapters III and IV for justifications, and for study of the accuracy of this approximation.) In a separable approximation,

$$t_s(p, k; s) = - g(p)\, g(k)/D(s). \tag{1.2}$$

The denominator function, $D(s)$ is expressed in terms of integrals involving the "form factor" $g(p)$ (see Chapter III). We consider in Chapter IV a rank-N separable t-matrix

$$t_N(p, k; s) = \sum_{i=1}^{N} \sum_{j=1}^{N} -g_i(p) \, \Delta_{ij}(s) \, g_j(k) \,. \tag{1.3}$$

We wish to emphasize at this point that we are approximating the t-matrix by expression (1.2) or (1.3); we *do not* assert that a corresponding separable approximation is accurate for the momentum-space potential $v(p, k)$ (see Chapter III).

After this digression for notation, we return to the Gordian knot, as it affects the two-nucleon problem. A major source of difficulty is the absence of a free neutron target. (Also neutron sources are less convenient experimentally than proton sources.) Thus neutron-neutron scattering measurements do not exist, except for a measurement of the scattering length by radiative capture of negative pions by deuterons. Neutron-proton scattering is measured less accurately than proton-proton scattering. Should we try to measure the t-matrix for neutron-neutron scattering by using the neutron in a deuteron as an almost free neutron target? It is clear that we must understand the three-nucleon problem to give a logically satisfying interpretation of $n - d$ scattering. (Perhaps the approximation of quasi-free scattering is satisfactory; but how can we be sure?)

Of course we could use the *assumption* of charge symmetry to derive the neutron-neutron phase shifts from measurements of the nuclear phase shifts in proton-proton scattering. But this assumption seems to fail, by a small amount, according to current calculations of the Coulomb energy difference ^{3}He$-^3$H. *If* the calculated Coulomb energy agreed with experiment, when we assume charge symmetry, we could draw the logical conclusion that both calculations and charge symmetry are valid. Now we can draw no completely firm conclusions.

As seen above, we have serious difficulties in finding the on-shell values of the two-nucleon t-matrix, when neutrons are involved. We have still more serious problems in extrapolating from on-shell values to the off-shell values needed in the three-nucleon problem (Levinger, 1973). The assumption of a local potential is a specific prescription for making this extrapolation; the assumption of a separable t-matrix is another; and there are many others. The assumption of locality is justified by custom; and to some extent by OPEP. The assumption of separability is justified by convenience in three-nucleon calculations. Neither justification is adequate.

1.4 Nucleon-Nucleon Wave Function

I can rephrase the problem discussed in the section above in simpler
terms, namely that of finding the eigenket $|E\rangle$ for the nucleon-nucleon
system at a specified energy E. Our general knowledge of quantum
mechanics leads us to the belief that complete knowledge of the
system of two-nucleons is given by knowledge of the wavefunction $|E\rangle$
at all energies. This belief is made precise by the remark that any
operator, such as the two-body Hamiltonian **H** can be expressed in
terms of its set of eigenvalues and eigenfunctions:

$$\mathbf{H} = \Sigma \, |E\rangle E \langle E| \, . \tag{1.4}$$

(The sum includes discrete values at negative energy, if any, and goes
over all positive energies.) Certainly knowledge of the Hamiltonian
should give us complete information on the $N - N$ system, equivalent
to that given by the off-shell t-matrix. For instance, we could in
principle solve the LS equation with the Hamiltonian 1.4. A more
practical method is given by Baranger et al. (1969), discussed in
Chapter III. Namely, use the wavefunction $|E\rangle$ to determine the half-
off-shell t-matrix; and then determine the completely off-shell t-matrix
using dispersion theory.

   We can look at the eigenfunction $|E\rangle$, for a specified partial wave,
in the coordinate representation. Its values at large r gives us the
measured phase shift. But in general the wavefunction remains unknown
at small r. Thus we cannot evaluate the Hamiltonian by 1.4. This lack
of knowledge of the wavefunction at small r is equivalent 'to the
uncertainty in extrapolation of the t-matrix from on-shell to off-shell
values. It is also equivalent to the uncertainty whether the Hamiltonian
in 1.4 corresponds to a local or a non-local potential.

   The wavefunction $|B\rangle$ for the single bound state, the deuteron at
$E = -B$, should be much easier to measure than the eigenfunctions at
positive energy. One usual method to measure a bound state eigenfunction is to scatter electrons from the bound state, and thus determine
its form factor. However, the deuteron problem is difficult, since the
deuteron has spin one. It therefore has three form factors, corresponding
to its three non-zero expectation values of charge (or monopole moment),
magnetic dipole moment and electric quadrupole moment. Experiments
to date have separated the magnetic form factor from a combination of
the two electric form factors; but the monopole and quadrupole form
factors remain entangled. I discuss in Chapter II an experiment on
polarization effects in electron-deuteron elastic scattering (Gourdin,
1966; Levinger, 1973) which should give considerably improved

knowledge of the deuteron wave function. This experiment is tangled in the Gordian knot, because its analysis involves: 1. knowledge of the neutron electromagnetic form factor; 2. knowledge of meson exchange effects, and effects of isobar admixture in the deuteron wave function.

1.5 Trinucleon

The great progress in calculation of the triton energy E_T for an *assumed* two-body potential has had three main causes: Faddeev's reformulation of the three-body problem; new formalism to solve the Faddeev equations; and new computing facilities and calculational techniques. We now have reasonable agreement (to an MeV) among all methods used for the same potential, and in many cases excellent agreement (to within 0.2 MeV).

We return to strands of the Gordian knot, when we compare calculated and experimental values of E_T. Besides our uncertainties in the on-shell two-body t-matrix discussed above, we have three additional uncertainties: 1. extrapolation of the two-body t-matrix to needed off-shell values; 2. relativistic effects; and 3. three-body forces. How shall we disentangle them? We also must be wary of drawing conclusions from the agreement (or disagreement) of calculation and experiment for a single number E_T.

We therefore compare calculated and experimental *functions*, namely the electromagnetic form factors for electron-trinucleon scattering. Since the trinucleon has spin and isospin each $\frac{1}{2}$, we have four different form factors. The magnetic isovector form factor is clearly affected by meson current or isobar effects: we have known for 25 years that the static magnetic moments of ^{3}H and ^{3}He lie outside the Schmidt lines (for a pure S state) and that any reasonable admixture of other states *increases* the discrepancy between calculation and experiment (Sachs, 1953). This large isovector effect could well be due to a one-pion exchange current, which could not affect the deuteron's (isoscalar) magnetic moment. Of course, we also need knowledge of the neutron electromagnetic form factor, to interpret electron-trinucleon scattering.

A "good" theory of particles and the forces between them would cut the Gordian knot: e.g., we could calculate the three-nucleon force with the same reliability that we now calculate the long-range part (OPEP) of the two-nucleon force. In the phenomenological approach followed here, we can only pick a little at various strands, making plausible assumptions as to which effects are large, and which can be neglected for the time being.

1.6 Summary

I have already surveyed the material in the following chapters. I repeat, more systematically, and quite briefly. In the next chapter, I discuss the two-nucleon system, emphasizing present gaps in our knowledge, and possible ways of closing those gaps. In Chapter III, I quote relevant formulas involving the two-nucleon t-matrix, and apply these formulas to the case of a separable central two-nucleon potential. I introduce the unitary pole approximation (UPA), and examine its accuracy. In the following chapter, I consider more complicated cases: rank-N separable central potentials, and also tensor separable potentials. I examine their accuracy in approximating the t-matrix for local potentials: e.g., Reid soft core singlet and triplet potentials.

In Chapter V, I present the Faddeev approach to the (simplified) model of a central spin-independent separable potential. I state the equations we need to solve for the trinucleon ground state for a tensor separable potential, and give some of the trinucleon results for the Reid soft core potential. In the following chapter, I outline five other methods now being used to solve the trinucleon: variational calculations; solution of two-dimensional integral equations; use of hyperspherical harmonics; t-matrix perturbation theory and Brueckner-Bethe theory. I compare results of different calculations of E_T for the Reid potential with each other, and with experiment. I also discuss calculations of the trinucleon electromagnetic form factors.

In the final chapter, I review briefly calculations of nucleon-deuteron scattering, by solution of the Faddeev equations. I then outline a program for some future work: better calculations on scattering, particularly inelastic nucleon-deuteron scattering; application of three-body calculations to models of heavier nuclei; and solution of the N-body problem, for N of four or more. I return to my provisional approach for untieing a portion of the Gordian knot.

II. A Critique of Experimental Knowledge of the Two-Nucleon System

2.1 Introduction

This chapter should be titled "Physical information needed to solve the three-nucleon problem". But Noyes (1969) has already used this title for a fine review. An equivalent approach would be "What we can learn about the two-nucleon system by studying the three-nucleon system". Both titles imply a lack of complete knowledge of the two-nucleon system particularly in the lower partial waves that dominate in the three-nucleon problem. Rather little firm knowledge is derived from theory; the analysis is primarily phenomenological. The earliest field theory calculation on this problem – the one-pion-exchange potential, abbreviated OPEP – has survived as the potential for relatively large nucleon-nucleon distance. (I present some evidence below for the validity of OPEP, after a summary of experimental data on the two-nucleon system.)

Only two-nucleon systems will be treated in this chapter: i.e., we shall not at this point try to use data on three-nucleon systems. We will allow non-strongly interacting particles as probes (e.g., electrons); but we omit here problems such as nucleon-deuteron or pion-deuteron scattering, treated briefly in Chapter VII.

There are a number of excellent reviews on nucleon-nucleon scattering (Noyes, 1972A; Signell, 1972; Moravcsik, 1972; Noyes, 1969; Breit, 1967; Signell, 1969; Moravcsik, 1963; and Wilson, 1963. See Mor 72 for a thorough literature review). I can, therefore, limit myself to a brief summary of experimental results, and their interpretation using phase parameters.

In order of increasing difficulty, experimentalists can measure the nucleon-nucleon total cross section, differential cross section, polarization, and triple scattering parameters. All these measurements have been made with good accuracy for the proton-proton system, and with only moderate accuracy for the neutron-proton system. There is only one direct measurement for the neutron-neutron system. (Of course, the absence of a free neutron target, and the greater difficulty of working with neutron beams as compared with proton beams, create these sizeable gaps in our knowledge.)

Scattering measurements can be performed up to very high energy in the center of momentum (CMS) system: e.g., with clashed beams at CERN for measurements of the proton-proton differential cross section. But "nuclear physicists" limit themselves to CMS energies not much higher than the 135 MeV needed for pion production: say 400 MeV in the laboratory system. Why? 1. The phase parameters become complex at energies above the threshold for pion production. 2. If we fit nucleon-nucleon scattering with a potential, then that potential would also be complex at high energies. 3. We *hope* we can calculate nuclear properties without using mesons explicitly. 4. It's really a definition of "nuclear physicist" that we leave meson production to the "high energy physicists". (I note parenthetically that the first and second objections are, strictly speaking, applicable even at low energies, due to nucleon-nucleon bremsstrahlung. In practice, the cross section for bremsstrahlung production is so small that it can be neglected.)

Measurements of nucleon-nucleon scattering are generally fitted using the Stapp, or barred, phase parameters. For the spin singlet state these are identical with the ordinary phase shift. The tensor force couples together two triplet states (e.g., for $J = 1$ with positive parity) giving us three phase parameters: $\bar{\delta}(^3S_1)$, $\bar{\delta}(^3D_1)$ and the mixing parameter $\bar{\varepsilon}_1$. For the purposes of application to the three nucleon system, we pay special attention to these four phase parameters: the phase shift for the $^1S_0$ state, and the three parameters listed above for the coupled states.

The phase shift for the $^1S_0$ state is measured with excellent accuracy in proton-proton scattering. (Of course, the effects of the Coulomb force must be removed.) These phase shifts can be extracted with poorer accuracy from neutron-proton scattering.

The considerable statistical analysis involved in extracting the phase parameters from the scattering data was, until recently, concentrated at Yale (SeaF 68) and at Livermore (MacA 69). Their general features are as follows (Mor 72). The $^1S_0$ phase shift starts at $0°$, increases but does not reach $90°$, and has a node at a proton laboratory energy near 260 MeV. $\bar{\delta}(^3S_1)$ starts at $180°$, since there is a bound deuteron, and falls monotonically, possibly having a node near 400 MeV. $\bar{\delta}(^3D_1)$ starts at zero, and decreases monotonically, reaching about $-25°$ at 300 MeV; while $\bar{\varepsilon}_1$ increases from zero to about $5°$ at 300 MeV. At present the relative uncertainties in $\bar{\varepsilon}_1$ are larger than those for the other phase parameters, (Sig 72).

The effective range parameters (scattering length a and effective range r) are extracted from the phase parameters at low energy. Noyes (1972 A) gives values in fair agreement with charge independence for the spin-singlet effective range parameters: an infinite scattering length and an effective range of 2.8 fm. Noyes indicates the following choices:

neutron-neutron: $a_{nn} = -17 \pm 1$ fm, $r_{nn} = 2.84 \pm 0.03$ fm

neutron-proton (spin singlet): $a_{np} = -23.715 \pm 0.015$ fm

$$r_{np} = 2.73 \pm 0.03 \text{ fm}$$

proton-proton: $a_{pp} = -7.824 \pm 0.005$ fm, $r_{pp} = 2.796 \pm 0.008$ fm .   (2.1)

The spin triplet bound state, the deuteron, has the accurately determined energy $-B = -2.22464 \pm 0.00005$ MeV. The effective range parameters are, scattering length 5.421 ± 0.007 fm, and effective range $\varrho(-B, 0) = 1.75 \pm 0.01$ fm.

The quadrupole moment of the deuteron, Q, is found from the extremely accurate measurements of the quadrupole coupling of the deuteron to the electric field gradient in a given molecule. As further calculations on this field gradient are made, the value of Q has crept up from early values of 0.274 fm^2 to 0.280 fm^2 (used in most current work, including mine below) to the very recent 0.2875 ± 0.002 fm^2 by Reid et al. (1972).

We can study the deuteron wave function $|B\rangle$ by elastic scattering in just the same way as atomic wave functions are investigated by x-ray scattering. The most useful probe for the deuteron is an electron beam of energy of order several hundred MeV. (We could also study elastic scattering of mu mesons, photons, or neutrinos. Inelastic electron scattering, or absorption of real photons, probe both the deuteron ground state and the excited final state.)

Elastic $e-d$ scattering involves both a deuteron form factor $\langle B|\exp(i\boldsymbol{q}\cdot\boldsymbol{r}/2)|B\rangle$ and isoscalar nucleon electric (E) and magnetic (M) form factors, G_{ES} and G_M (Glendenning et al., 1962; Wilson et al., 1964; Gourdin, 1966). It is hard to determine both deuteron and nucleon properties in the same experiment: indeed, experiments to date have given rather little information about either quantity. A serious complication is that the deuteron has *three* form factors: the electric monopole F_0, magnetic dipole F_1, and electric quadrupole F_2. (Their static limits give the three possible moments of a system of unit spin: the charge, the magnetic dipole moment, and the electric quadrupole moment.) The differential cross section (Glendenning, 1962) is proportional to $(F_0^2 + F_2^2)\,G_{ES}^2$, and a second term involving magnetic effects. The electric and magnetic parts are separated by experimentalists using the measured angular dependence at fixed momentum transfer. But the combination $(F_0^2 + F_2^2)$ is insensitive to the deuteron wave function. We return later in this chapter to $e-d$ scattering.

2.2 Meson Theory

Noyes (1969) has summarized the success of OPEP in fitting nucleon-nucleon data, particularly its use by Grashin (1959) and by Moravcsik (1963) for calculation of phase parameters for partial waves with high orbital angular momentum. But OPEP clearly fails to tell the whole story for lower partial waves. How accurate is OPEP for low spin at some specified nucleon-nucleon distance r? (Feshbach et al., 1967, 1968; Lomon, 1970.) How should we generalize OPEP to find a theoretical nucleon-nucleon interaction at smaller values of r? The two questions are closely related. Two different, competing, extensions of OPEP are the use of two-pion-exchange-potentials (TPEP) and of one-boson-exchange-potentials (OBEP). The TPEP generally suffers from problems in treating nucleon-antinucleon pairs, and in treating meson-meson interaction. On the other hand, the OBEP suffers from uncertainties as to which bosons are "really there" — e.g., should we use a 0^+ boson called σ at 300–400 MeV? (Lasinski et al., 1973.) Also, are we allowed to treat a massive composite system, which decays very rapidly (such as the ϱ resonance) as a single exchanged particle?

There has recently been considerable progress in developing a meson theory of nuclear forces that combines the best features of TPEP and OBEP. (Sig 72) I shall confine my qualitative exposition to the work of the Paris group. (CotL 73; VinR 73. See CheD 72 for a somewhat different approach.) First I note that we should expect a meson theory of nuclear forces to be based on experimental data concerning mesons. Thus, going back to the analogous problem of the Coulomb force calculated in quantum electrodynamics, the values of the electron charge e and the photon rest mass are taken from experiment. So must the vector character of the electric field. (There are calculations of the fine structure constant; but they are controversial, to say the least.) The electron charge can be determined from the cross section for elastic electron-photon scattering (Thomson cross section). The value of zero for the photon rest mass can be found from the independence of the speed of electromagnetic waves in vacuum on their frequency.

OPEP is based on analogous experiments. The meson-nucleon coupling constant is found (to much lower accuracy than the electron-photon coupling constant) from measurements on pion-nucleon elastic scattering. The meson mass is measured in a mass spectrometer. The pseudoscalar character of the pion is determined from measurements of pion interaction, such as the Panofsky ratio.

Once we introduce states with two pions, we need to use experiments on pion-pion scattering, as a function of energy, and of the spin and isospin of the two-pion system. If a certain phase shift passes rapidly through 90°, this corresponds to a boson resonance used in OBEP; but non-resonant two-pion states can be included also. Unfortunately current experiments on pion-pion scattering generally involve 3-hadron systems: e.g., the reaction $\pi + N \rightarrow \pi + \pi + N$. Besides difficulties of interpretation, there are also sizeable experimental errors. Vinh-Mau et al. (1973) consider two sets of pion-pion phase shifts: I simplify by selecting the set (with no S-wave pion-pion resonance) that gives the best overall fit to nucleon-nucleon scattering. The Mandelstam representation is used to predict the nucleon-nucleon scattering amplitude, given the pion-nucleon and pion-pion scattering amplitudes.

Three-pion states are treated, following OBEP, as the ω resonance, with the $\omega - N$ coupling determined from other experiments: the isoscalar electromagnetic nucleon form factors, and production of the ω-resonance by electron-positron collision. I show in Fig. 2.1 one set of phase shifts ($^1D_2$) calculated by the Paris group. The agreement illustrated between theory and experiment is satisfactory, and clearly a vast improvement over the use of OPEP.

Unfortunately the Paris group has not yet completed work on the lowest partial waves which we need for trinucleon calculations. Also,

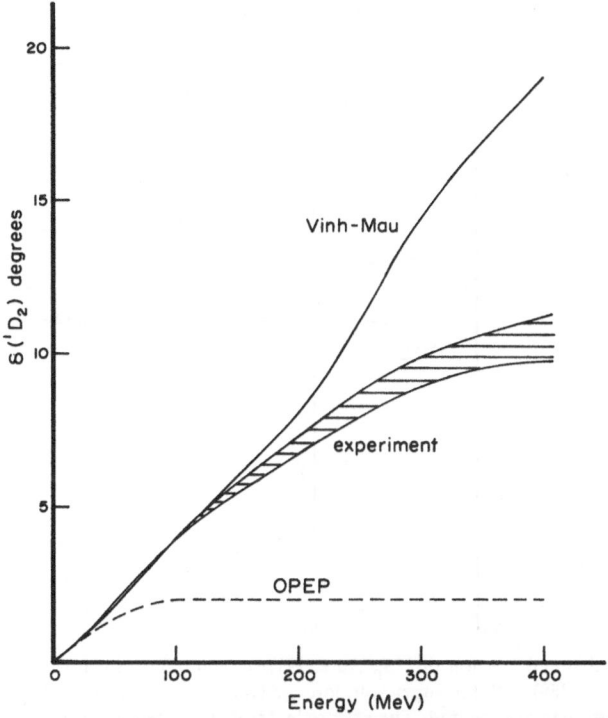

Fig. 2.1. Phase shifts for the $^1D_2$ state, calculated by Vinh-Mau et al. (1973) shown as a solid line, compared with those for the one-pion-exchange-potential (dashed line) and experimental values, shown shaded bars. The ordinate is degrees, the abscissa laboratory energy in MeV

we need off-shell values of the t-matrix for the trinucleon; rather than the on-shell values of the nucleon-nucleon t-matrix shown in Fig. 2.1. The work of the Paris group might be pursued to satisfy these two needs. But for present I must still limit myself to phenomenological nucleon potentials.

2.3 Phenomenological Potentials

We can reach a qualitative conclusion from TPEP, OBEP or the recent calculations. First, OPEP is particularly good for the *tensor* potential at moderate distances (say greater than 3 fermis), but less satisfactory for a central (spin triplet or singlet) potential. Why? The pion has no competition in producing a long range tensor force; but it has very strong competition for moderately long range central forces, by exchange of a σ resonance, or of two pions in a 0^+ state. The TPEP range is shorter

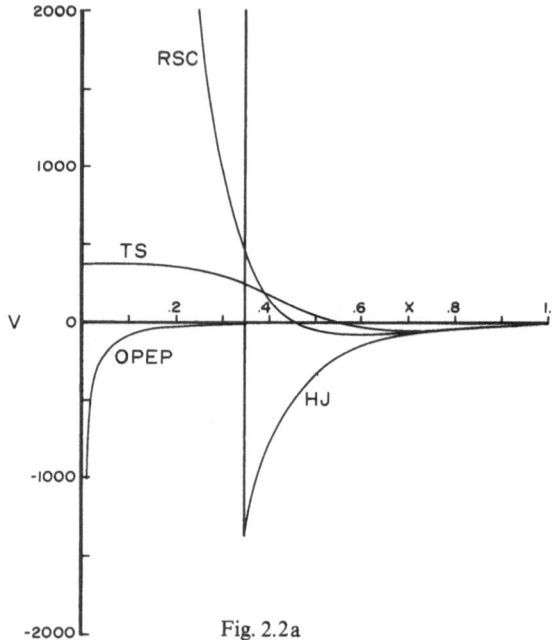

Fig. 2.2a

Fig. 2.2. Local potentials for the $^1S_0$ state: HJ for Hamada-Johnston (1962), RSC for Reid soft core (1968), TS for super soft core, deTourreil and Sprung (1973), and OPEP for one-pion-exchange potential. The energy in MeV is plotted against nucleon-nucleon distance x

than OPEP, but the coupling is much larger for TPEP. This qualitative conclusion is supported by phenomenological fits shown in Fig. 2.2. Vinh-Mau et al. find similar results for their potential.

I show in Fig. 2.2 three different phenomenological fits to the central potential acting in the ^1S state: Hamada-Johnston hard-core (1962), Reid soft core (1968) and deTourreil-Sprung super-soft core (1973). Each potential approaches the OPEP value at large r; but the approach is seen to be rather slow. Below energies are in MeV, r in fm and $x = \mu_\pi r = 0.70\,r$.

$$V_{HJ} = V_{OPEP}[1 + 8.7\exp(-x)/x + 10.6\exp(-2x)/x^2], \ x > 0.343$$

$$V_{HT} = \infty, \ x < 0.343$$

$$V_{RSC} = V_{OPEP} - 1650.6\ \exp(-4x)/x + 6484.2\exp(-7x)/x$$

$$V_{dTS} = 374.98\exp(-4.4224\,r^4) + \{-1001.6\exp(-3.6071\,x)/x$$
$$+ V_{OPEP}\}\,\{1 - \exp(-r^4)\}$$

$$V_{OPEP} = -10.463\exp(-x)/x\,.$$

(2.2)

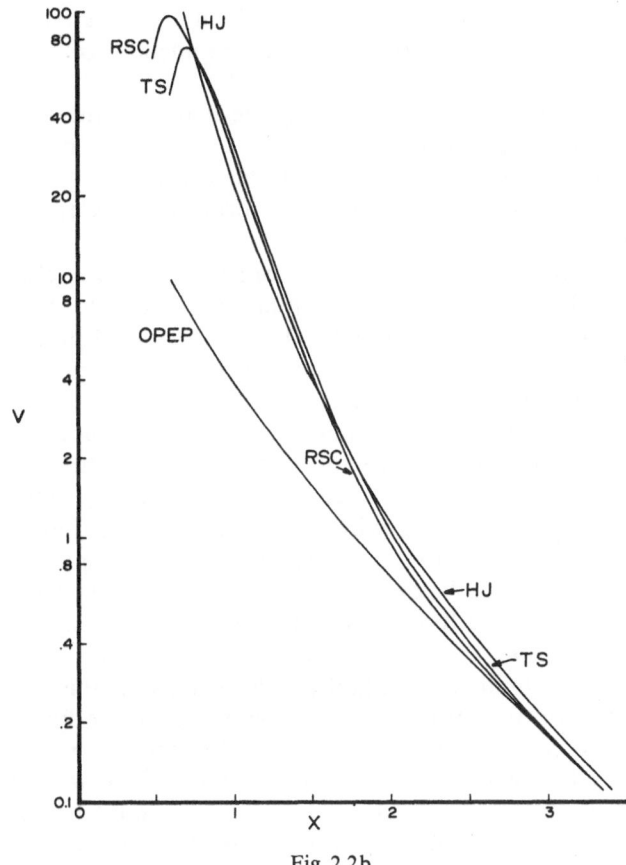

Fig. 2.2b

Near the origin (within $\frac{1}{2}$ fm) the potentials are obviously going to be very different from each other, as indicated by the names "hard core", "soft-core" and "super-soft core". It is not so obvious a priori but is quite clear from the figure — that the potentials also disagree greatly in the region from 1 to 2 fm. The choice of behavior near the origin causes corresponding drastic differences in behavior at "moderate distances". These differences may seem surprising in view of the Gelfand-Levitan theorem that *exact* knowledge of the phase shift at *all* energies uniquely determines an (assumed local and energy-independent) potential. The three potentials fit the accurate measurements with reasonable values of χ^2, up to a laboratory energy of some 300 MeV. The apparent violation of the uniqueness theorem arises from the experimental errors, and from the limitation to 300 MeV (so that the potential is real).

So even if the potential were local, it is not well determined even for the experimentally favorable ^1S state, determined in accurate proton-proton scattering.

Further, all bets are off concerning the *nature* of the potential at small values of r. Present field-theoretical calculations suggest appreciable non-locality (Signell, 1972). On the other hand, if we use a phenomenological potential we have no firm basis (except habit) for assuming its locality.

We must be careful in our use of the word "local". A *completely* local and energy-independent central potential $V(r)$ would be identical for any value of angular momentum L. This form of local potential died almost 40 years ago, when Majorana introduced a space-exchange operator, so the potential has a part which changes sign depending on whether L is odd or even. The next relaxation of standards for "an acceptable local potential" came some 15 years ago with the use of a spin-orbit $L \cdot S$ force. Since this force didn't solve the difficulty in trying to fit ^1S and ^1D phase shifts with the same potential (Noyes, 1969), further non-locality was introduced: either a quadratic spin-orbit force (Hamada-Johnston, 1962), or a force with depends explicitly on L^2 (de Tourreil, 1973), or a force that is different for each combination of J, L, and S (Reid, 1968).

The Reid potential (for example) is local only in the *weak* sense that for a specified partial wave (or coupled waves) it is a function only of the nucleon-nucleon distance. (I shall, for brevity, use the term "local" below to mean local in this weak sense.) Once we have given up some strong form of locality, why should we want to keep weak locality? Occam's razor ("Do not multiply hypotheses needlessly.") does favor keeping weak locality. Also, the potential should approach the local OPEP at large inter-nucleon distances; so maybe it is also local at smaller distances. A decision concerning the local character of the potential cannot be made in our present phenomenological framework, and is not decided conclusively in present field-theoretic work.

The same basic question, which we keep returning to, enters in the extrapolation of the nucleon-nucleon t-matrix from its measured on-shell values to the off-shell values needed in the three-nucleon problem.

We can get some ways off the energy shell in the spin-triplet case by study of the deuteron wave function $|B\rangle$. Two unresolved problems are: 1. What is the value of the deuteron percent D-state, p_D? 2. Is there a "hole" in the S-wave radial function $u(r)$ at small r, corresponding to a strong, very short-range nucleon-nucleon repulsion?

The value of p_D has intrinsic interest, and it is particularly important for us here, as it is a main source of uncertainty in calculations on the three-nucleon system. A low value of p_D is associated with a weak, very long range tensor force (Biedenharn, 1958) and a correspondingly strong central force. (The two forces must be adjusted to give the observed values of B, effective range, and Q.) The tensor force is relatively ineffective in binding the three-nucleon system. But a strong central force is quite effective, giving a large negative value of triton energy E_T. Thus we should expect E_T to be a monotonic increasing function of p_D: cf. Phillips (1968) and Brady et al. (1969).

Attempts to obtain an accurate value of p_D from precise measurements of the magnetic moments of proton, neutron, and deuteron are frustrated by the likely existence of appreciable contributions to the deuteron's magnetic moment from 1. relativistic effects (Blatt-Weisskopf, 1953); 2. effects of the spin-orbit or other momentum-dependent terms in the nucleon-nucleon potential (Feshbach, 1957; Mukherjee et al., 1973) 3. meson exchange effects (Adler, 1966); and 4. admixture of nucleon isobars in the deuteron (Arenhövel et al., 1972). Each effect is hard to estimate (sometimes even the sign is in doubt) and we have the additional difficulty that is unclear (to me, at least) whether each effect should be evaluated separately, or whether they "overlap", and we should omit some effects listed to avoid "double counting". Current estimates of these effects suggest that they should not be large enough to move p_D "very far" from its value of 4%, found from the magnetic moment: perhaps $p_D = (4 \pm 3)\%$.

Phenomenological potentials that do *not* impose the constraint of long range OPEP behavior, give a large range of values for p_D: e.g., Brady et al. (1969) find $0.78\% \lesssim p_D \lesssim 7\%$; Mongan (1969) finds $p_D \simeq 1\%$; and Biedenharn (1958) finds $3\% \lesssim p_D \lesssim 12\%$. It would be helpful to have firm limits for the allowed range of p_D. I (Levinger, 1969) made a simple but unrealistic estimate of the minimum value, but I have not been able to find an upper bound.

We approximate the quadrupole moment Q by

$$Q = (1/\sqrt{50}) \int_0^\infty r^2 u(r) w(r) \, dr. \tag{2.3}$$

We use the asymptotic value for the S wave $u(r)$ from effective range theory:

$$u(r) = \sqrt{2\gamma} (1 - \gamma \varrho)^{-\frac{1}{2}} \exp(-\gamma r). \tag{2.4}$$

Here the effective range $\varrho(-B, -B)$ is taken as 1.76 fm, B is the deuteron binding energy, and $\gamma = B^{\frac{1}{2}} = 0.232 \text{ fm}^{-1}$. We want to minimize

$$p_D = \int_0^\infty w^2(r) \, dr, \tag{2.5}$$

for the known $u(r)$, with the D wave $w(r)$ chosen so that Eq. (2.3), gives the experimental value of Q. The Schwartz inequality shows that we should choose $w(r)$ proportional to $r^2 u(r)$. Choosing the proportionality constant so that (2.3) gives us Q, and evaluating (2.5), we find

$$(p_D)_{min} \equiv p_m = (25/12)(1 - \gamma\varrho)(4\gamma^2 Q)^2 = 0.44\%. \tag{2.6}$$

As I remarked earlier (1969), the proportionality of $w(r)$ to $r^2 u(r)$ is unrealistic; since it corresponds to a long range force. Klarsfeld at Orsay (private communication) has made more realistic estimates of p_m, using the restriction that for a short range force $w(r)$ must approach the form for a free D wave, namely $\exp(-\gamma r)(1 + 3/\gamma r + 3/\gamma^2 r^2)$. He constrains $w(r)$ to agree exactly with this free D wave for $r > R$ and finds that varying $w(r)$ for $0 < r < R$ increases p_m to a value that depends monotonically on R. Klarsfeld obtains $p_m \simeq 3\%$ for $R \simeq 1/2\gamma \simeq 2$ fm. Klarsfeld obtains a similar value of p_m assuming that $w(r)$ approaches the value for an OPEP potential.

Presumably Mongan (1969) achieves his low value of p_D by use of a wave function which does not correspond to an asymptotic OPEP potential. Burnap et al. (1970) studied the simpler case of Brady's wave functions for separable potentials of Yamaguchi form with different values of p_D. The analytical wave function is substituted into the Schrödinger equation to find the corresponding (local) central and tensor potentials, which are then compared with OPEP. Burnap finds poor agreement for $0.78\% \lesssim p_D \lesssim 2\%$, and fair agreement for $4 \lesssim p_D \lesssim 7\%$. We can understand Burnap's first result by making an approximation to his Eq. (10) for large r:

$$V_T \simeq Cr \exp[(\gamma - \beta_t) r] \tag{2.7}$$

Here C is a constant, and β_t is the range parameter in Yamaguchi's tensor form factor, $T(p) = -tp^2/(p^2 + \beta_t^2)^2$. Numerical values for the monotonic increasing function $\beta_t(p_D)$ are given by Brady et al. (1969). The small value $p_D = 0.78\%$ gives the small value $\beta_t = 0.51$ fm^{-1}, for which 2.7 gives a tensor potential V_T proportional to $r \exp(-0.28 r)$. This behavior is *very* different from the OPEP proportionality to $r^{-1} \exp(-\mu r)(1 + 3/\mu r + 3/\mu^2 r^2)$, with $\mu = 0.70$ fm^{-1} from the pion mass.

At present I copy, without great conviction, the range for p_D usually quoted: $3\% \lesssim p_D \lesssim 7\%$.

If the potential were local and energy-independent and if there is a node in the phase parameter $\bar{\delta}(^3S_1)$ then there must be a strong repulsion at a distance of order $\frac{1}{2}$ fermi. We might have a hard core, or a soft core, or even a "super-soft" core (de Tourreil et al., 1973). Alternatively, we might use a boundary condition model (Partovi, 1972). Any of these four cases would have a "hole" in the wave function of radius roughly $\frac{1}{2}$ fermi.

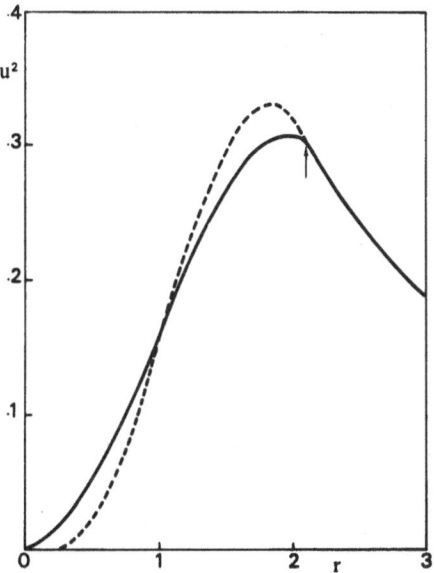

Fig. 2.3. The solid line gives the squared radial wave function for the velocity-dependent potential, 2.8. The arrow shows the edge of the square well, where there is a discontinuity in the slope of the wave functions. The dashed line gives the squared wave function for a static potential with repulsive core. The two wave functions are identical for $r \gtrsim 2.0\,\mathrm{fm}$

But this hole may be absent for a non-local potential: e.g., the velocity-dependent potential of Razavy et al. (1962), used by Levinger et al. (1961). Figure 2.3 illustrates the squared wave function $u^2(r)$ for a local square well with a hard core, compared to that for a potential quadratic in the momentum. Razavy used the form

$$v(r, p) = -V_0 J_1(r) + (\lambda/M)\boldsymbol{p} \cdot J_2(r)\,\boldsymbol{p} \tag{2.8}$$

with both functions J_1 and J_2 chosen as square wells of the same range.

Alternatively, we can follow Haftel's (1973) procedure of constructing a family of phase-equivalent potentials by the use of a short-range unitary transformation, $\mathbf{U}$. That is, we choose the transformation subject to the constraint.

$$\langle r|U|r'\rangle \underset{r\,\text{or}\,r' \to \infty}{\longrightarrow} \delta(\boldsymbol{r} - \boldsymbol{r}') \tag{2.9}$$

The Hamiltonian $\mathbf{H}$ is transformed into a different Hamiltonian

$$\tilde{\mathbf{H}} = \mathbf{U}\mathbf{H}\mathbf{U}^\dagger . \tag{2.10}$$

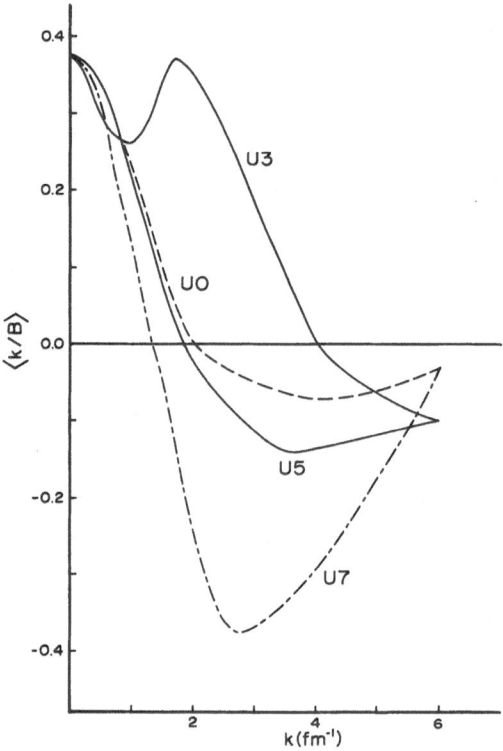

Fig. 2.4. The "deuteron" momentum space wave function for a local potential (curve U 0), and for non-local phase-equivalent potentials U 3, U 5 and U 7. From Haftel (1973)

The transformed Hamiltonian in general has a non-local potential. The eigenstates $|\psi_n\rangle$ of **H** are transformed into

$$|\tilde{\psi}_n\rangle = \mathbf{U}|\psi_n\rangle .$$ (2.11)

Since the unitary transformation has short range, the transformed and original eigenstates $|\tilde{\psi}_n\rangle$ and $|\psi_n\rangle$ have identical behavior at large r: hence they have the same phase shifts for positive energy and the same eigenvalues for bound states. But the eigenfunction for a bound state is in general changed. I copy as Fig. 2.4 Haftel's figure for the momentum space wavefunction for the original Hamiltonian (U 0) and for three transformed Hamiltonians, using unitary operators he calls U 3, U 5 and U 7.

We can discuss this same question in t-matrix language: given the on-shell values of the t-matrix, how much freedom do we have in the residue of the t-matrix at the deuteron pole? (Fiedeldey, 1969; Haftel, 1973.) As shown in the next chapter, this residue is proportional to the deuteron momentum space wave function, which in turn is related to the existence of a hole in $u(r)$, and to the shape of the deuteron form factor for electron scattering.

2.4 Electron-Deuteron Scattering

How can we determine experimentally if the deuteron wave function has a hole like that shown as the dotted curve in Fig. 2.3? It turns out (Brady et al., 1972) that this question and that posed above concerning the value of p_D can both (in principle) be answered by measurements of the deuteron form factors for elastic electron scattering. We define monopole and quadrupole form factors,

$$F_0(q^2) = \int_0^\infty (u^2 + w^2) j_0(qr/2)\, dr\,, \tag{2.12}$$

$$F_2(q^2) = \int_0^\infty 2w(u - w/8^{\frac{1}{2}}) j_2(qr/2)\, dr\,, \tag{2.13}$$

where j_0 and j_2 are spherical Bessel functions. Since the nucleons have structure, the amplitude for electron scattering involves the products $G_{ES}(q^2)\,F_0(q^2)$, and $G_{ES}(q^2)\,F_2(q^2)$ where $G_{ES}(q^2)$ is the isoscalar electric nucleon form factor. (There is also magnetic scattering, but as discussed above this is poorly understood even in the static limit, $q^2 = 0$. We will make an analysis in which we do not need to know the amplitude for magnetic scattering. Of course, we are not certain that we can neglect the poorly understood exchange and isobar terms in electric scattering; but at least we know from charge conservation that they are zero for $F_0(q^2)$ in the static limit. Calculations by Adler (1966) and by Williams (1971) suggest that these extra electric terms may not be of great importance for the moderate values of momentum transfer considered below. Also see Arenhövel (1972) and Überall (1971).)

Measurements of the differential cross section for elastic electron-deuteron scattering (Elias et al., 1969; Galster et al., 1971) give the quantity (Glendenning et al., 1962; Gourdin, 1966; Wilson et al., 1964).

$$R \equiv (d\sigma/d\Omega)/\sigma_{NS} = G_{ES}^2(F_0^2 + F_2^2) + G_{MS}^2 F_{mag}^2(1 + 2\tan^2 \theta/2)\,. \tag{2.14}$$

Here θ is the electron scattering angle in the laboratory system, and we have neglected terms of order $q^2/4M_D^2$, where M_D is the deuteron mass. σ_{NS} is the "no structure" differential cross section, and G_{MS} and F_{mag} are the nucleon magnetic and deuteron magnetic form factors respectively. Extrapolation of R, for fixed momentum transfer q, to $\tan^2 \theta/2 = -\frac{1}{2}$ eliminates magnetic scattering, giving us the combination $G_{ES}^2(F_0^2 + F_2^2)$ for electric scattering. Unfortunately this combination is insensitive to the two features we want to study. First, at values of q of order $5\,\mathrm{fm}^{-1}$ where the monopole form factor F_0 may have a node, F_2 is relatively large, so the effect of the monopole diffraction minimum is small. Second, at low values of momentum transfer, $q \simeq 1\,\mathrm{fm}^{-1}$, the quadrupole form factor $F_2(q)$ is indeed sensitive to the value of p_D, but the quadrupole form factor is much less than the monopole form factor, so the differential cross section is insensitive to the value of p_D.

Gourdin (1964, 1966) suggested unscrambling the monopole and quadrupole form factors by measuring the tensor polarization of the deuteron in (e, d) scattering. Our proposal below (Levinger, 1973A; Brady et al., 1972) is a scheme to eliminate also the small effect of magnetic scattering. It is a matter of experimental convenience whether

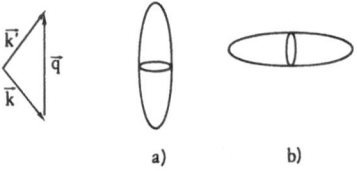

a)   b)

Fig. 2.5. The momentum transfer is q. In the first sketch, the deuteron spin is aligned with the momentum transfer; in the second sketch the two vectors are perpendicular

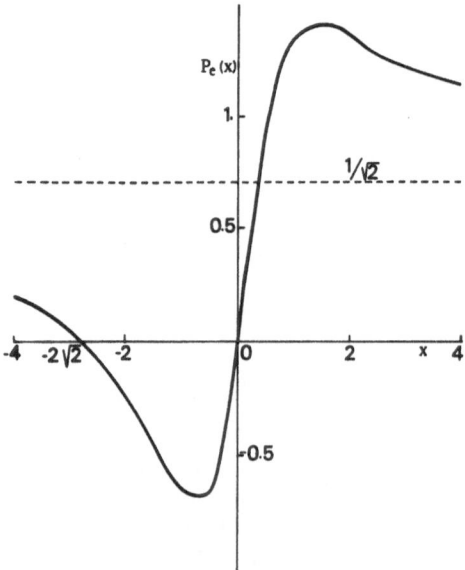

Fig. 2.6. Plot of the extrapolated polarization P_e, vs. $x =$ quadrupole form factor/monopole form factor. See Eq. (2.17)

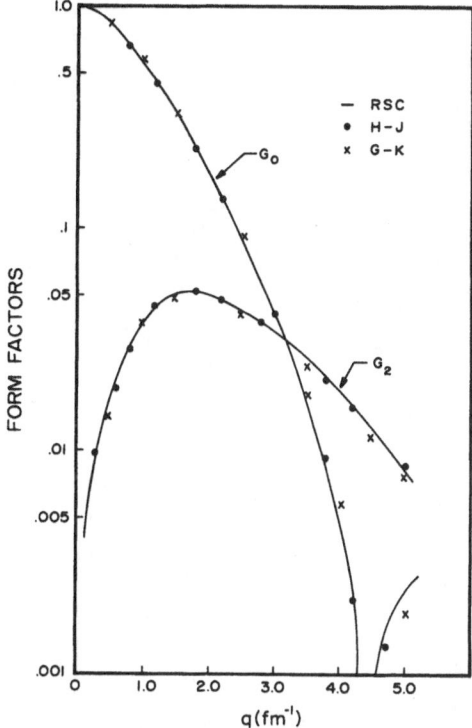

Fig. 2.7. Semi-logarithmic plot of absolute value of deuteron form factors vs. momentum transfer, for three potentials with cores: Reid soft core, Hamada-Johnston, and Glendenning-Kramer. For the definitions of monopole G_0 and quadrupole G_2, see Eqs. (2.12) and (2.13) and text below. From Brady et al. (1972)

one measures the polarization of the recoil deuteron (Ohlsen, 1972) or instead measures electron scattering from aligned deuterons. Time reversal invariance shows that the two measurements give the same information. I shall discuss the latter, since it appeals to my physical intuition. Consider scattering shown in Fig. 2.5 for two different polarizations, $m = 0$ and $m = 1$. Here m gives the component of deuteron spin along the direction of momentum transfer q. (The state $m = -1$ gives the same scattering as $m = 1$.) The Fourier transform $\langle B | \exp(iq \cdot r/2 | B \rangle$ will be different for the two cases illustrated. We define tensor polarization $P = T^{20}$, using the Madison convention. (Barschall et al., 1971; Brady et al., 1972.) Translating Gourdin's equations into our notation, or by an independent calculation, we have

$$P(\theta) = [2 G_0 G_2 + G_2^2/\sqrt{2} - (G_{\text{mag}}^2/\sqrt{8})(1 + 2\tan^2\theta/2)]\, C(q) \qquad (2.15)$$

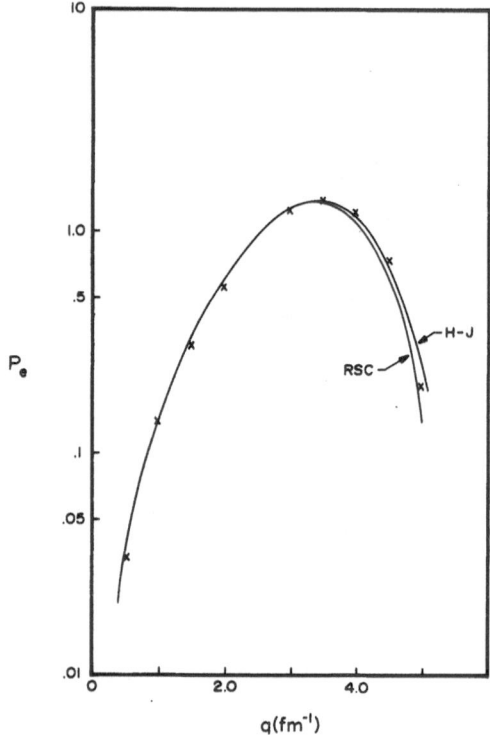

Fig. 2.8. Extrapolated polarization P_e for Reid soft core and for Hamada-Johnston deuteron wave functions: vs. momentum transfer. From Brady et al. (1972)

where

$$[C(q)]^{-1} = G_0^2 + G_0^2 + (1 + 2 \tan^2 \theta/2) G_{\text{mag}}^2 . \tag{2.16}$$

If we extrapolate the polarization at fixed q to an angle such that $\tan^2 \theta/2 = -\frac{1}{2}$, we can again eliminate the magnetic contribution. We obtain the extrapolated $P_e(x)$ given by

$$P_e(x) = (2x + x^2/\sqrt{2})/(1 + x^2) . \tag{2.17}$$

We note that the polarization depends only on the ratio

$$x \equiv G_2/G_0 = F_2/F_0 . \tag{2.18}$$

That is, polarization measurements are of no use in the determination of nucleon electromagnetic form factors; but we do not need to know G_{ES} to calculate the polarization.

Figure 2.6 illustrates the function $P_e(x)$. Of course, $P_e(x)$ is proportional to x for small values of x; P_e increases to a maximum value of 1.4 for

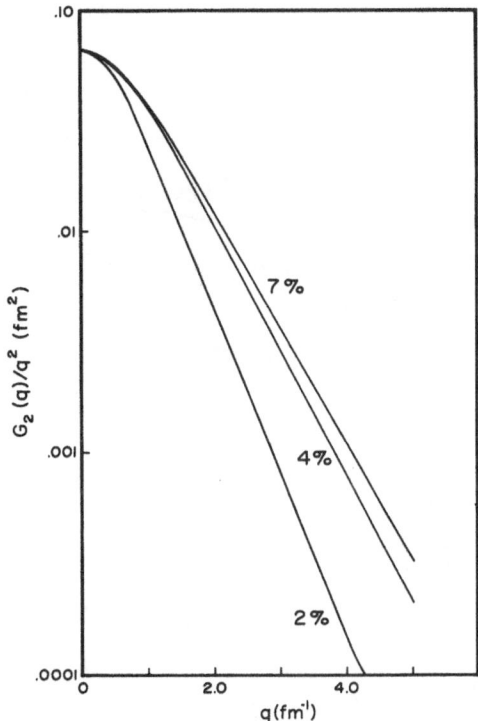

Fig. 2.9. Semi-logarithmic plot of quadrupole form factor/q^2 for Yamaguchi shape potentials, with varying percentage of deuteron D-state. From Brady et al. (1972)

x near 1.5; and then falls to $1/\sqrt{2}$ as x increases to infinity. P_e has a node at the negative value, $x = -\sqrt{8}$.

We consider first three "realistic potentials" with $5\% \leqq p_D \leqq 7\%$, with a core at small r, and a corresponding hole in $u(r)$. 1. Glendenning (1962) has published F_0 and F_2; we use his potential 8. Brady et al. (1972) evaluated G_0 and G_2 for the 2. Reid soft core and 3. Hamada-Johnston wave functions. Figure 2.7 shows the diffraction minimum in $G_0(q)$ at $q \simeq 4.4\,\text{fm}^{-1}$. As q increases from zero, x increases from zero to infinity (at the diffraction minimum in G_0), jumps to $-\infty$, and then increases again. Figure 2.7 shows that the polarization $P_e[x(q)]$ first increases, and then falls rapidly as the diffraction minimum is approached. Figure 2.8 shows this effect, but does not go to high enough values of q to show the sign change of the polarization. The recent de Tourreil-Sprung (1973) deuteron wave functions give almost the same shape of $P_e(q)$.

We contrast this shape for the polarization with that for Yamaguchi wave functions, for different values of p_D. (Brady, 1969; Levinger, 1973 A.)

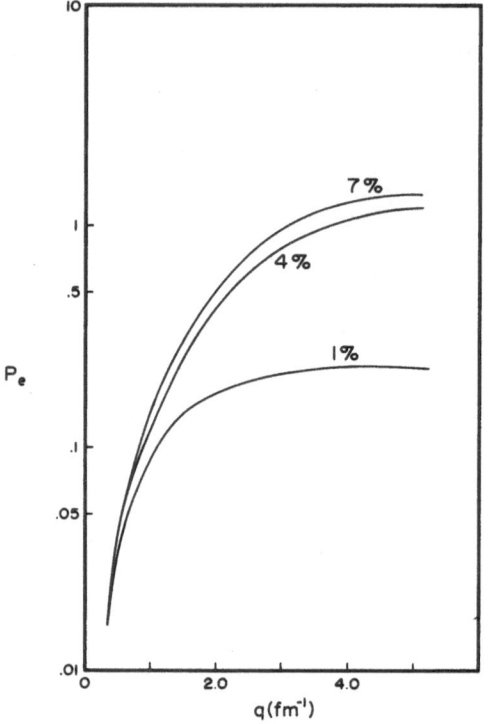

Fig. 2.10. Semi-logarithmic plot of extrapolated polarization vs. momentum transfer for Yamaguchi shape wave functions, with varying percentages of deuteron D-state. From Brady et al. (1972)

The wave functions are chosen to give identical quadrupole moments, so from the definition 2.13 for $F_2(q)$ the ratio $F_2(q)/q^2$ must approach the same value as q approaches zero, namely $\sqrt{2}\,Q/6$. (In *principle* measurements of polarization provide a means of measuring Q without knowledge of the gradient of the electric field strength in a molecule.) Despite the agreement as q approaches zero, there is considerable variation of $G_2(q)$ with p_D for $q \simeq 1$ fm^{-1}, as illustrated in Fig. 2.9. This dependence is easily understood as follows (Levinger, 1973A). We can fit Q with a low value of p_D by using a D-wave with a long tail; but this tail contributes little to $G_2(1)$ due to destructive interference. This variation in $G_2(q)$ produces a strong variation of polarization with p_D shown in Fig. 2.10. Note that for $0.5 \leqq q \leqq 2$ fm^{-1}, the polarization curve for a Yamaguchi wave function with 7% D-state lies close to that for the three realistic potentials illustrated in Fig. 2.8. We see that measurements of P at low q should give us the value of p_D; but very accurate measurements would

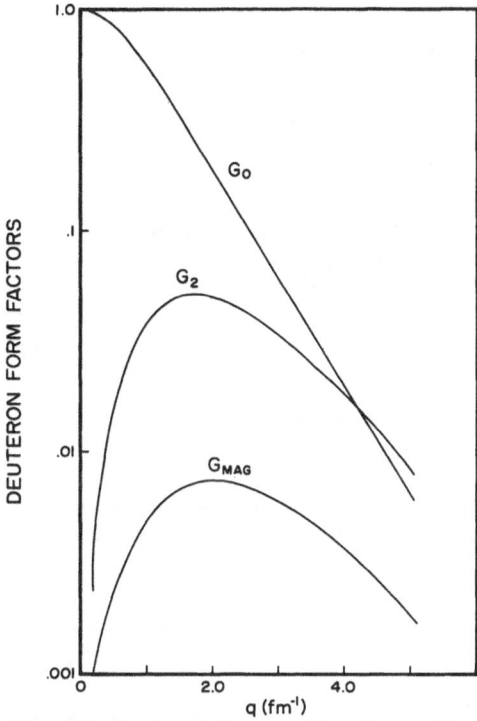

Fig. 2.11. Semi-logarithmic plot of deuteron form factors G_0 and G_2 vs. momentum transfer for Yamaguchi shape potential with 7% deuteron D-state. From Brady et al. (1972)

be needed to distinguish between D-state probabilities of 4% and 7%.

Figure 2.10 also shows that the polarization curves for Yamaguchi wave functions do *not* show the maximum exhibited in Fig. 2.8 for realistic wave functions. The absence of a maximum is due to the relatively slow decrease of $G_0(q)$ for the Yamaguchi wave function, illustrated in Fig. 2.11 for $p_D = 7\%$; note that G_0 does not have a node.

The nodeless character of G_0 for Hulthén-Yamaguchi wave function $u(r)$ is associated with its smooth character. (Thus a wave function with $u^2(r) = \delta(r - b)$ would have an infinite number of nodes in the form factor.) Figure 2.12 shows the results of Shapiro's calculation with the Levinger-Rojo wave functions (1961) shown in Fig. 2.1. (The form factors are in excellent agreement, not illustrated, for $0 \leq q \leq 2 \text{ fm}^{-1}$.) Figure 2.13 shows three other examples from Haftel (1973): a unitary transformation gives a nodeless monopole form factor.

The Hulthén-Yamaguchi choice for $u(r)$ for a separable potential is likely to be unrealistic, since this separable potential cannot repro-

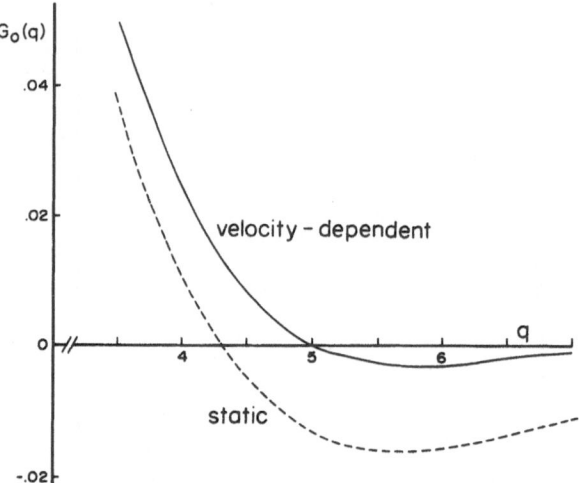

Fig. 2.12. Diffraction minimum for monopole form factor vs. momentum transfer for the velocity-dependent potential and static potential with hard core, with wave functions illustrated in Fig. 2.3

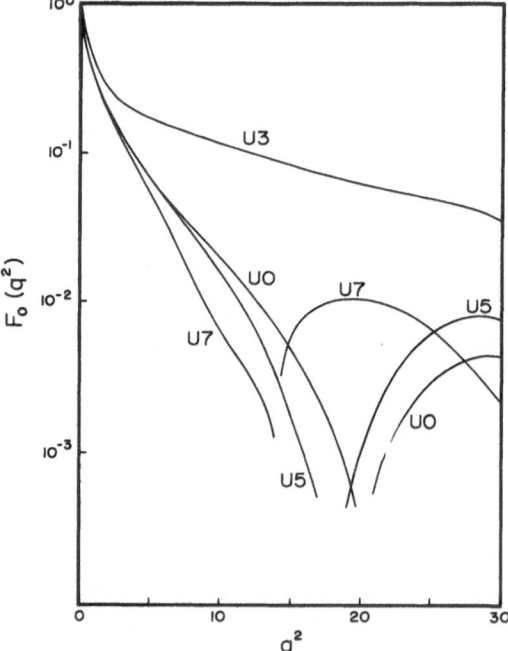

Fig. 2.13. Absolute value of monopole form factor $F_0(q^2)$, for Haftel's (1973) phase-equivalent potentials, with momentum space wave functions shown in Fig. 2.4. The squared momentum-transfer q^2 is given in fm^{-2}

duce the possible node in $\delta(^3S_1)$. But it is possible to use two-term separable potentials (Mongan, 1969) or potentials such as 2.8 or Haftel's unitary transformation which give the node in the phase shift without giving a node in the form factor. [Brady (1969) shows this for Mongan's potential.]

These proposed experiments on deuteron polarization effects demand an electron accelerator with energy at least of order $\frac{1}{2}$ GeV, to obtain a high enough value of q (of order $4\,\text{fm}^{-1}$) at not too large an electron scattering angle. The accelerator should have a very good duty cycle, so that background can be greatly reduced by measurements of electron-deuteron coincidences. These difficult experiments would provide significant information both on the value of p_D, and on the existence of a hole in the deuteron wave function.

2.5 Charge Symmetry

Another open question concerns the charge independence, and the charge symmetry of the two-nucleon potential. The mass difference between charged and neutral pions gives some charge dependence.

Evidence for (small) deviations from charge symmetry comes from the three-nucleon system: i.e., the comparison of the calculated Coulomb energy of ^{3}He with the experimental value (Fabre, 1972). We discuss Fabre's calculations of the Coulomb energy later in this review; we limit ourselves here to brief remarks on a theory of a non-symmetric term in the nucleon-nucleon force. The failure of charge symmetry comes from the failure of isospin to describe systems of exchanged particles.

First let's consider OPEP. The pion has a definite isospin, so the OPEP is charge symmetric. (This can be shown formally by writing OPEP in isospin notation. In a less formal manner, the coupling of a neutron to a neutral pion is the same in magnitude as the proton coupling to a neutral pion; so the neutron-neutron OPEP equals the proton-proton OPEP.) The same argument for charge symmetry holds for exchange of any mesonic system, provided that the exchanged system has a pure isospin, e.g., the ϱ resonance, which has the same isospin as the pion. The purity of isospin of a meson resonance can be tested experimentally by looking for isospin forbidden decays: e.g., three pion decay of the ϱ resonance; or two pion decay of the omega resonance. But the reaction $\omega^0 \rightarrow \pi^+ + \pi^-$ *does* occur with a branching ratio of $1.3 \pm 0.3\%$ (Lasinski, 1973). That is, there is a mixture of isospin zero and isospin one systems, for the vector (1^-) mesons ω^0 and ϱ^0.

Henley (1972) considers $(\varrho^0 - \omega^0)$ mixing, giving the diagram of Fig. 2.14 as a part of the nucleon-nucleon potential. This mixture of

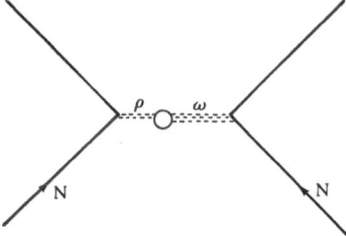

Fig. 2.14. Diagram for a contribution to the nucleon-nucleon potential that is not charge-symmetric, due to mixing of the ϱ^0 and ω^0 resonances. From Henley et al. (1972)

two meson resonances of the same angular momentum and parity, but with different isospin, is likely to be especially large for this example, due to their approximate degeneracy. Gourdin (1970) analysed experiments on production of pion pairs by clashed electron-positron beams (the ω decay into 2π) to estimate the mixing as 6%. Henley (1972) and van Wageningen (1972) use Gourdin's estimate to find the breaking of charge symmetry: they conclude that the neutron-neutron force is about $\frac{1}{2}$% stronger than the nuclear part of the proton-proton force, in rough agreement with Fabre's result.

2.6 Summary

Many workers have fitted the phase parameters and the static deuteron properties with local or non-local potentials. We shall not attempt even a partial list. A given potential fit provides a prescription to extrapolate from the measured on-shell t-matrix to the off-shell values we shall need below for the three-nucleon system. For instance, the Reid (1968) potential is (weakly) local, with a soft core. The recent de Tourreil-Sprung (1973) potential is also local, with a super-soft core. Riihimaki (1970) uses a long range local potential combined with a short range separable potential of second rank. On the other hand, Mongan (1969) uses a two-term separable potential. The first three workers impose the constraint that the potential approach OPEP at large r, while Mongan ignores this constraint.

I will not try to give an argument here for, or against, any of these or other phenomenological parametrizations. Instead I shall concentrate below (Chapters 5 and 6) on a different question. Suppose we *assume* a given potential, such as the Reid soft core. Do we have reliable methods to calculate the properties of the three-nucleon system for this potential? If we can answer "yes", then (and only then) it would be worthwhile to consider a variety of local and non-local potentials to learn the sensitivity of three-nucleon properties to the assumed two-nucleon t-matrix. I have chosen the Reid soft core because it has the general features that are likely to survive for a two-nucleon potential (long-range OPEP behavior, tensor force, soft or supersoft core) and because it has been used recently in calculations of the three-nucleon system by a variety of different calculational techniques. Also the Reid "soft core" represents in fact a very strong short range repulsion, that provides serious numerical problems even in the solution of the two-nucleon problem, and a fortiori in the three-nucleon problem. So use of Reid's soft core provides a severe test of our calculational ability in the three-nucleon problem.

III. Two-Body Scattering

3.1 Introduction

In this chapter I first review scattering theory (Messiah, 1961) for the two-nucleon system: the relation between the t-matrix and the scattering amplitude, and the Low and Lippmann-Schwinger equations for the t-matrix. I present the equations for a partial wave for central forces, using the normalization of Goldberger et al. (1964) and Watson et al. (1967). The Low equation shows that the t-matrix is factorable or separable (Lovelace, 1964) very near its pole at negative energy (i.e., near a bound state). I then introduce the unitary pole approximation or UPA (Fuda, 1968A) to construct a separable unitary t-matrix corresponding to a specified wave function for the bound state. (See Watson et al. for earlier use of a factorable approximation.) After a general discussion concerning possible uses of the UPA, two examples are presented (Harms, 1970) in some detail: a Yukawa potential, and the Reid singlet soft core potential (Reid, 1968).

3.2 Scattering Amplitude

The on-shell elements of the t-matrix $\langle p | t(k^2 + i\varepsilon) | k \rangle$, with $p = k$ and ε a negligible positive quantity, are proportional to the amplitude $f(\theta)$ for elastic scattering from a state with asymptotic solution the plane wave $|k\rangle$ to another plane-wave state $|p\rangle$. The proportionality constant between $f(\theta)$ and on-shell elements of the t-matrix is different in different references, because of different normalizations used for the plane-wave states. Goldberger et al. (1964) use

$$\langle p | k \rangle = \delta(p - k) \tag{3.1}$$

where the right side is a 3-dimensional Dirac delta function. Then a plane wave has a spatial representation of

$$\langle r|k\rangle = (\sqrt{2\pi})^3 \exp(ik \cdot r).$$ (3.2)

We have the completeness relation for the unit matrix, $\mathbf{I}$,

$$\mathbf{I} = \int d^3 q |q\rangle\langle q|.$$ (3.3)

In this normalization we have

$$f(\theta) = -2\pi^2 \langle p|v|\psi_k\rangle = -2\pi^2 \langle p|t(k^2 + i\varepsilon)|k\rangle.$$ (3.4)

Here $|\psi_k\rangle$ is the exact solution of the Schrödinger equation with potential v. $\langle r|\psi_k\rangle$ agrees asymptotically with $\langle r|k\rangle$ at large r for a short-range potential. We use units with $\hbar$ = nucleon mass = 1. The expression using the matrix element of the potential is proved from scattering theory (Messiah, 1962); the expression using the t-matrix constitutes a definition of the t-matrix.

The ket $|k\rangle$ is a solution of the Schrödinger equation for kinetic energy operator $\mathbf{H}_0$ with a corresponding free particle Green's function

$$\mathbf{G}_0(s) = \mathbf{I}/(s - \mathbf{H}_0).$$ (3.5)

(Bold face roman designate a matrix or operator.) On the other hand, $|\psi_k\rangle$ is the solution of the Schrödinger equation for the complete Hamiltonian $\mathbf{H} = \mathbf{H}_0 + \mathbf{v}$. Messiah uses these statements, and the relation (3.4) between matrix elements of the potential and of the t-matrix, to derive the Lippmann-Schwinger (LS) equation

$$\mathbf{t}(s) = \mathbf{v} + \mathbf{v}\,\mathbf{G}_0(s)\,\mathbf{t}(s).$$ (3.6)

In (3.5) and (3.6) the energy s is any complex number. For on-shell elements, we choose $s = k^2 + i\varepsilon$ as in (3.4); but (3.6) allows us to go off-shell using any value of s. (See 1.1 for the definition of "off-shell", and see the accompanying text for an introduction to the physical interpretation of off-shell elements of the t-matrix.)

We also follow Messiah in introducing the Green's function $\mathbf{G}(s)$ for the complete Hamiltonian $\mathbf{H}$

$$\mathbf{G}(s) = \mathbf{I}/(s - \mathbf{H}).$$ (3.7)

We can then rewrite the Schrödinger equation as the Low equation

$$\mathbf{t}(s) = \mathbf{v} + \mathbf{v}\,\mathbf{G}(s)\,\mathbf{v}.$$ (3.8)

We combine Eqs. (3.3), (3.5), and (3.6) to obtain the LS equation in the momentum representation

$$\langle p|t(s)|k\rangle = \langle p|v|k\rangle + \int d^3 q \langle p|v|q\rangle \langle q|t(s)|k\rangle/(s - q^2).$$ (3.9)

Here the potential in momentum space is found from (3.2)

$$\langle p|v|k\rangle = (2\pi)^{-3} \int v(r) \exp[i(k - p) \cdot r] \, d^3 r.$$ (3.10)

Note that if we neglect the second term on the right in (3.9), and use (3.4) and (3.10) we have the well known Born approximation to the scattering amplitude; an iterative solution of (3.9) gives the second Born approximation, and higher terms.

For central forces, we find it very useful to use a partial wave representation. For example, we use S-waves, and calculate the matrix element of the operator equation (3.8) between two states each of angular momentum zero, and momenta of magnitude p and k respectively. We find

$$\langle p, 0|t(s)|k, 0\rangle \equiv t(p, k, s) = v(p, k) + \int_0^\infty q^2 \, dq v(p, q) \, t(q, k; s)/(s - q^2)$$ (3.11)

where the S-wave momentum space potential

$$\langle p, 0|v|k, 0\rangle \equiv v(p, k) = (2/\pi) \int_0^\infty r^2 \, dr j_0(pr) \, v(r) j_0(kr).$$ (3.12)

We extract the S-wave part of the complex scattering amplitude, (3.4), express it in terms of the phase shift δ, and in terms of on-shell elements of the t-matrix between s-waves

$$\exp(i\delta) \sin\delta/k = -2\pi^2 t(k, k; k^2 + i\varepsilon).$$ (3.13)

3.3 Separable t-Matrix

We follow Lovelace (1964) and Watson et al. (1967) in using the Low equation to show that the t-matrix is separable or factorable very near a pole at negative energy $-B$, with wave function $|B\rangle$. We assume that there is only one bound state: this simplifies notation and corresponds to the nucleon-nucleon system. We expand the unit matrix in (3.7) using the complete set consisting of $|B\rangle$ and the positive energy eigenfunctions $|E\rangle$. [The latter have normalization $\langle E'|E\rangle = \delta(E - E')$.] The Low equation (3.8) becomes

$$\langle p|t(s)|k\rangle = \langle p|v|k\rangle + \langle p|v|B\rangle(s + B)^{-1}\langle B|v|k\rangle$$
$$+ \int_0^\infty dE \langle p|v|E\rangle(s - E)^{-1}\langle E|v|k\rangle.$$ (3.14)

Near the pole at $s = -B$, the second term on the right increases without bound, while the first and third terms are bounded. (We neglect the special case of matrix elements $\langle p|v|B\rangle$ or $\langle B|v|k\rangle$ equalling zero.)

We then approximate the t-matrix by the second term

$$\langle p|t(s)|k\rangle \simeq \langle p|v|B\rangle \langle B|v|k\rangle/(s+B). \tag{3.15}$$

This argument works for central or for non-central forces (Fuda, 1968A). For central forces, consider the elements of the t-matrix for S-waves. We find by the same argument,

$$t(p,k;s) \simeq \langle p|v|B\rangle \langle B|v|k\rangle/(s+B) = g(p)\,g(k)/(s+B). \tag{3.16}$$

That is, we have obtained the separable form (1.2) as a good approximation near the bound state pole. Below we follow Fuda in developing the UPA, and obtain a better expression for the energy dependence of the t-matrix than that given by the denominator of (3.16). But first, we introduce the expression for the "form factor" $g(p)$ in (1.2), and use $\mathbf{v} = \mathbf{H} - \mathbf{H}_0$ with $\mathbf{H}|B\rangle = -B|B\rangle$ and $\langle p|\mathbf{H}_0 = p^2\langle p|$. Then,

$$g(p) = \langle p|v|B\rangle = -(p^2 + B)\langle p|B\rangle. \tag{3.17}$$

We are now $\frac{2}{3}$ of the way to the UPA (i.e., we have the P and the A); but we don't have a unitary expression for the t-matrix. Fuda obtained a unitary expression by assuming that we have a real but non-local separable potential. (The t-matrix for any real potential conserves probability, or is unitary.)

We first discuss non-local potentials and summarize Yamaguchi's treatment (1954) of a separable central potential. [A separable potential had been introduced some twenty years earlier by Wigner. See Blatt and Weisskopf (1952), Chapter III, Eq. (3.25). Yamaguchi's work was independent, and initiated modern interest in this approximation. Also the Yamaguchi's (1954) treated separable tensor forces for the first time.]

3.4 Non-Local Potential

We implicitly assumed above that we were dealing with a *local* potential $v(r)$; however the formalism also works for a *non-local* potential. Physicists almost always use local potentials, and must therefore apologize when they use non-local (or velocity dependent) potentials. By a local potential we mean that the result of operating with the potential v on a wave function ψ is a new function which is represented in coordinate space by

$$\langle r|V|\psi\rangle = V(r)\,\psi(r). \tag{3.18}$$

That is, a local potential is a very simple operator in coordinate space: it merely multiplies the wave function at a given position by a function of that same position. Any other potential is called "non-local". We replace the right side of (3.18) by an integral over vector coordinate r', and replace the function $V(r)$ of a single vector by a function $W(r, r')$ of two space vectors, and integrate over all r' giving

$$\langle r | V | \psi \rangle = \int W(r, r') \, \psi(r') \, \mathrm{d}^3 r' . \tag{3.19}$$

We can reduce (3.19) back to the local case (3.18) with a clever choice of the operator W, namely,

$$\langle r | W | r' \rangle = W(r, r') = V(r) \, \delta(r - r') \tag{3.20}$$

Substitution of (3.20) into (3.19) verifies that we have recovered the desired local form. With this example we can interpret a general potential function $W(r, r')$ as follows. We may replace the infinitely narrow peak of the Dirac delta function in (3.20) by a peaked function of finite width b'. The integral in (3.19) then utilizes $\psi(r')$ for r' different from r by a length of order b' to produce the result of operating on the wave function with the non-local potential. If b' goes to zero, as in (3.20) we have a local potential, where we need know only the value of the wave function at *precisely* the coordinate r; but for a non-local potential we need to know the wave function in a sphere of radius order b' centered at r.

The form (3.19) for a non-local potential is sufficiently general to include a momentum-dependent potential (Razavy et al., 1962) or a Majorana exchange potential that changes sign depending on the even or odd value of the orbital angular momentum (Blatt and Weisskopf, 1952).

The potential $W(r, r')$ used above is non-local, and independent of energy. We could instead extend the concept of an energy-independent local potential illustrated in (3.18) to give an energy-dependent local potential: we need only assume that the potential function $V(r)$ depends on the energy E of the system. Or we could consider the still more complicated non-local energy-dependent potential by allowing the function $W(r, r')$ to depend on the energy. In either case of energy-dependence the operator $v(s)$ in the LS or Low equations (3.6 or 3.8) would have an explicit dependence on energy.

I return below under "epistemological remarks" to the problem of determining whether the nucleon-nucleon potential is local or non-local, and whether it depends on energy or is energy-independent. Since we do not yet have experimental evidence to decide this question, I make the arbitrary choice in this book of emphasizing non-local but energy-independent potentials. Below, a potential is assumed to be independent of the energy, unless I specify to the contrary.

3.5 Separable Potential

A *separable* potential (of *rank-one*, which we use in this chapter) replaces the local form (3.20) with

$$\langle r \,|\, W \,|\, r' \rangle = W(r, r') = -\lambda f_1(r)\, f_2(r') \tag{3.21}$$

I assume that λ is a constant, independent of energy; positive λ gives an attractive potential. The condition that the potential is Hermitian gives us

$$\langle r \,|\, W \,|\, r' \rangle = \langle r' \,|\, W \,|\, r \rangle^* = -\lambda^* f_1^*(r')\, f_2^*(r)$$
$$= -\lambda f_1(r)\, f_2(r') . \tag{3.22}$$

Then the functions f_1 and f_2 can be chosen real and equal. We denote this function by f, and use a real positive quantity λ as the strength of the attractive potential (Blatt and Weisskopf, 1952, Chapter III, Eq. (3.25); Yamaguchi, 1954)

$$\langle r \,|\, W \,|\, r' \rangle = -\lambda f(r)\, f(r') . \tag{3.23}$$

(Comparing (3.20) and (3.23), we see that there is one case for which a local potential is separable: namely $V(r) = \delta(r)$ in (3.20) and $f(r) = \delta(r)$ in (3.23). We shall return to this delta function potential for illustration; but we want to examine potentials of finite range and plausible shape.)

The separable form (3.23) unfortunately does not approach a local potential such as OPEP at large r. Consider a central potential of range of order b, so that $f(r')$ is small if $r' \gg b$. Suppose r/b is large; but the integral in (3.19) has its major contribution from $r' \simeq b$, or very different from r. Contrast this with the local potential (3.20) where $r' = r$. Indeed, we find that a separable potential has the paradoxical property that the larger r, the larger the *difference* between r' and r for the important contributions to the integral in (3.19). That is, the separable potential becomes "more and more non-local" as r becomes larger.

The reader may well wonder why I continue to discuss separable potentials in view of this strong argument against their plausibility. My excuse is to reiterate that the assumption of separable potentials is of heuristic value, since it leads to a separable approximation to the *t*-matrix which is frequently useful.

(One can also argue that separable potentials are so much easier to work with that they should be used for pedagogical reasons in the presentation of quantum mechanics, whether they are reasonable or not. Thus, we find below that the differential equation for the two-body problem for a local central force is reduced to quadrature in the separable case. In general, the separable approximation removes one variable from the problem; thus reducing the difficulty of the problem by at least an order of magnitude).

It proves convenient to work with the spatial Schrödinger equation for local potentials since the potential is a simple operator (3.18). But (Yamaguchi, 1954) the Schrödinger equation for separable potentials is easier to solve in momentum space. In general the momentum space wave function is a solution of

$$p^2 \phi(\mathbf{p}) + \int v(\mathbf{p}, \mathbf{k}) \, \phi(\mathbf{k}) \, \mathrm{d}^3 k = E \phi(\mathbf{p}). \tag{3.24}$$

An S-wave momentum space wave function $\phi(p)$ satisfies the integral equation

$$p^2 \phi(p) + \int v(p, k) \, \phi(k) \, k^2 \mathrm{d} k = E \phi(p). \tag{3.25}$$

(Note that twice the reduced mass equals the nucleon mass of unity.) Here the "potential in momentum space" $v(p, k)$ is given by a Fourier transformation of $W(r, r') = -\lambda f(r) f(r')$, for a central separable potential

$$v(p, k) = -\lambda(2\pi)^{-3} \int_0^\infty \exp(i\mathbf{p} \cdot \mathbf{r}) \, f(r) \, \mathrm{d}^3 r \, \exp(-i\mathbf{k} \cdot \mathbf{r}) \, f(r') \, \mathrm{d}^3 r'$$
$$= -\lambda g(p) \, g(k), \tag{3.26}$$

where the function $g(p)$ is the Fourier-Bessel transform of the function $f(r)$. I absorb a factor $\sqrt{2/\pi}$ in $g(p)$: thus,

$$g(p) = \sqrt{2/\pi} \int_0^\infty j_0(pr) \, f(r) \, r^2 \mathrm{d} r. \tag{3.27}$$

Substituting (3.26) into the Schrödinger equation (3.25), we find that $\phi(p)$ satisfies the *algebraic* equation

$$p^2 \phi(p) - \lambda g(p) \int g(k) \, \phi(k) \, k^2 \mathrm{d} k = E \phi(p). \tag{3.28}$$

We solve (3.28) for an attractive separable potential for a bound state at negative energy $E = -B = -\gamma^2$. We replace the integral in (3.28) by an unknown *number*, A

$$A \equiv \int_0^\infty g(k) \, \phi(k) \, k^2 \mathrm{d} k. \tag{3.29}$$

Substituting in (3.28) we easily solve for the momentum space wave function $\phi(p)$. Of course we obtain the same result as (3.17) for the relation between the form factor and the momentum space wave function of the bound state,

$$\phi(p) = \lambda A g(p)/(\gamma^2 + p^2). \tag{3.30}$$

We can easily go back and forth from knowledge of the form factor $g(p)$ to the wave function $\phi(p)$ or vice versa. We omit normalization of $\phi(p)$, and instead study the determination of the function $\gamma(\lambda)$: i.e., given the shape of the form factor $g(p)$, how does the energy depend on the strength λ of the potential? We determine $\lambda(\gamma)$ by substituting the wave function (3.30) into the definition (3.29).

$$A = \int_0^\infty g(k)\,\lambda A g(k)\,k^2\,\mathrm{d}k/(\gamma^2 + k^2),$$
or
$$\lambda(\gamma) = \left[\int_0^\infty g^2(k)\,k^2\,\mathrm{d}k/(\gamma^2 + k^2)\right]^{-1}. \tag{3.31}$$

Yamaguchi (1954) showed from (3.31) that a separable potential produces at most one bound state. Since $\mathrm{d}\lambda/\mathrm{d}\gamma$ is positive, $\lambda(\gamma)$ is a monotonic increasing function of γ. The minimum value λ_m, for zero binding, is

$$\lambda_m = [\int g^2(k)\,\mathrm{d}k]^{-1}. \tag{3.32}$$

For $\lambda < \lambda_m$ the separable potential is not strong enough to produce a bound state. (Of course this property of a minimum strength for a potential of specified shape to produce a bound state in three dimensions holds also for local potentials.) If the potential has strength $\lambda_1 > \lambda_m$, there is a *unique* value λ_1 satisfying (3.31).

Since a physicist's experience with a local potential leads him to expect the possibility of several bound states for a strong attractive potential, our result that an arbitrarily strong but separable potential gives a single bound state seems peculiar. I therefore give a second mathematical proof, in hopes of making the result above seem convincing, as well as correct.

Assume that there were two bound states: at $-B = -\gamma^2$ and at $-C = -\delta^2$. The momentum space eigenfunction $\eta(p)$ for the latter would be found from $\phi(p)$ in 3.30, merely by changing γ to δ. But then the orthogonality integral could not be zero, by inspection.

$$\int_0^\infty \phi(p)\,\eta(p)\,p^2\,\mathrm{d}p = \int_0^\infty g^2(p)\,(\gamma^2 + p^2)^{-1}\,(\delta^2 + p^2)^{-1}\,p^2\,\mathrm{d}p \neq 0. \tag{3.33}$$

(I omit constants in the integral.) Since the orthogonality property must hold even if the potential is non-local (provided that it is energy independent), we conclude that we *cannot* have two bound states for a separable potential.

Another peculiarity of a separable potential of the form (3.26) is that it acts only in a two body S-state (Yamaguchi, 1954). The integral in (3.24)

$$\int v(p, k)\,\phi(k)\,\mathrm{d}^3 k = -\lambda g(p) \int g(k)\,\phi(k)\,\mathrm{d}^3 k \tag{3.34}$$

selects out just the spherically symmetric part of the wavefunction $\phi(k)$. Of course, one could also choose a separable potential to act in some other specified partial wave.

These two peculiarities of a separable potential of form (3.26) are shared by a local potential $\delta(r)$: it has a single bound state, and it acts in an S state.

The LS equation (3.11) for the generalized partial wave t-matrix is also solved easily by introducing a separable potential of form (3.26).

$$t(p, k; s) = -\lambda g(p) g(k) - \lambda g(p) \int_0^\infty g(q) (s - q^2)^{-1} t(q, k; s) q^2 \, dq . \qquad (3.35)$$

Define the function

$$A(k) = \int_0^\infty g(q) (s - q^2)^{-1} t(q, k; s) q^2 \, dq . \qquad (3.36)$$

Then

$$t(p, k; s) = -\lambda g(p) [g(k) + A(k)] . \qquad (3.37)$$

Substituting (3.37) in (3.36) we find that $A(k)$ is proportional to $g(k)$

$$A(k) = -\lambda g(k) \left[\int_0^\infty g^2(q) q^2 (s - q^2)^{-1} dq \right] \Big/ \left[1 + \lambda \int_0^\infty g^2(q) q^2 (s - q^2)^{-1} dq \right] . \qquad (3.38)$$

Substituting in (3.37) we find the t-matrix

$$t(p, k; s) = -g(p) g(k)/D(s) \qquad (3.39)$$

where the denominator function

$$D(s) = \lambda^{-1} + \int_0^\infty g^2(q) (s - q^2)^{-1} q^2 \, dq . \qquad (3.40)$$

Equation (3.39) shows that a separable potential (3.26) gives a completely separable t-matrix; the dependence of t on the two momenta and one energy factor separates into three functions each of a single variable. The t-matrix (3.39), with choice (3.17) for the form factor, gives the unitary pole approximation.

If there is a negative value of s, called $-B$, such that $D(-B) = 0$, then the t-matrix (3.40) has a pole at $s = -B$. The condition that $D(-B) = 0$ is just (3.31) found above from the Schrödinger equation in momentum space. We have illustrated the general result (3.16) that the t-matrix is singular and separable at a bound state pole. Our first calculation starting from (3.25) was just a solution of the homogeneous form of the LS equation $t(p, k; -B) = v(p, k) \mathbf{G}_0(-B) t(p, k; -B)$.

The denominator $D(s)$ and the t-matrix are real for negative energy s. We now consider $s = 0$, and positive s. In the first case, we relate the on-shell t-matrix to the scattering length a; in the second case to the phase shift δ.

Putting $p = k = s = 0$ in (3.39) and (3.40), we find the on-shell t-matrix, which equals the scattering length a within a proportionality constant.

$$a = 2\pi^2 \langle 0|t(0)|0\rangle = -2\pi^2 g^2(0)/\left[\lambda^{-1} - \int_0^\infty g^2(q)\,\mathrm{d}q\right]. \qquad (3.41)$$

We obtain a negative scattering length if $\lambda^{-1} > \int_0^\infty g^2(q)\,\mathrm{d}q = \lambda_m^{-1}$, corresponding to no bound state [cf. (3.32)]; an infinite scattering length if $\lambda^{-1} = \lambda_m^{-1}$ corresponding to a "bound state", at energy 0; and a positive scattering length for $\lambda^{-1} < \lambda_m^{-1}$ for a bound state.

Consider on-shell values at positive energy $s = k^2 + i\varepsilon$; here the t-matrix is complex due to the pole in the denominator function (3.40). We solve (3.13) for the real phase shift δ, using $p = k$ in (3.39).

$$\exp(i\delta)\sin\delta/k = 2\pi^2 g^2(k)/D(k^2 + i\varepsilon) \qquad (3.42)$$

or,

$$\tan\delta(k) = -\operatorname{Im} D(k^2 + i\varepsilon)/\operatorname{Re} D(k^2 + i\varepsilon). \qquad (3.43)$$

We find the real part, $\operatorname{Re} D$, by taking the Cauchy principal value P in (3.40). The imaginary part, $\operatorname{Im} D$, is proportional to its value at the pole, $q = k = s^{\frac{1}{2}}$

$$\operatorname{Im} D(k^2 + i\varepsilon) = -\pi k g^2(k)/2. \qquad (3.44)$$

Then the phase shift is given by

$$\tan\delta(k) = \tfrac{1}{2}\pi k g^2(k)\left[\lambda^{-1} + \mathrm{P}\int_0^\infty g^2(q)\,(k^2 - q^2)^{-1}\,q^2\,\mathrm{d}q\right]^{-1}. \qquad (3.45)$$

We are interested in making a separable approximation for an attractive potential, assumed weak enough that there is no bound state. Then for very small values of k the denomination in (3.45) is positive, and so is $\tan\delta(k)$. (We are just repeating our calculation on the scattering length.) Do we find any zeros in the phase shift as we increase k, besides the asymptotic approach to zero for infinite momentum? We can find examples of the form factor $g(k)$ such that (3.45) would have a zero in the denominator for $k = k_r$, and thus produce a change of sign of $\tan\delta(k)$. (It happens that Yamaguchi's choice for $g(k)$ give no such zeros.) But this method of producing a zero is objectionable on the physical grounds that it corresponds to putting a (non-observed) resonance in the continuum at k_r in the nucleon-nucleon system. Tabakin (1968) has tried, cleverly but unsuccessfully, to obtain this node in the phase shift without obtaining this resonance. His trick is to let $g(k_r) = 0$, so that $\tan \delta(k_r)$ should not go to infinity. Tabakin's attempt has been criticized by Bolsterli (1969) and by Brady et al. (1969) on different grounds: the definition of the scattering matrix; that the resonance could still appear in more complex nuclear reactions; and that Tabakin's t-matrix would be a very poor approximation to the t-matrix for a local potential.

Recently Osborn (1973) showed that a separable approximation to the t-matrix must disagree seriously with the exact t-matrix for a local potential. (His argument is for a rank N separable potential; but holds a fortiori for a rank-one separable potential considered here.) Osborn's argument is based on the non-compact character of a local potential, and its corresponding t-matrix; whereas a separable t-matrix is compact. The difference between them

$$\Delta\mathbf{t}(s) \equiv \mathbf{t}(s) - \mathbf{t}^u(s) \tag{3.46}$$

is then non-compact, and therefore not square-integrable. [Here $\mathbf{t}(s)$ is the exact and $\mathbf{t}^u(s)$ a separable t-matrix, both functions of energy s.] That is, the trace

$$\mathrm{tr}(\Delta t\, \Delta t^\dagger) = \int \langle p|\Delta t(s)|p'\rangle^2 \, \mathrm{d}^3 p \, \mathrm{d}^3 p' = \infty . \tag{3.47}$$

At first sight, this infinite mean square deviation between $\mathbf{t}(s)$ and $\mathbf{t}^u(s)$ is very disturbing: it is hard to believe that $\mathbf{t}^u(s)$ might be a satisfactory approximation to $\mathbf{t}(s)$. A solution (Levinger, 1973B; Sloan and Gray, 1973) of this paradox lies in finding the source of the divergence and examining what is meant by a satisfactory approximation.

The lack of convergence in (3.47) stems from the non-square integrability of a local potential. That is

$$\mathrm{Tr}(vv^\dagger) = \int \langle p|v|p'\rangle^2 \, \mathrm{d}^3 p \, \mathrm{d}^3 p' = \int [f(q)]^2 \, \mathrm{d}^3 q \int \mathrm{d}^3 Q = \infty . \tag{3.48}$$

Here $q = p - p'$ and $Q = p + p'$. The divergence arises from very large values of Q; and a closer analysis (Le 73; SloG 73) shows the same source for the divergence (3.47). But we introduce a separable approximation $\mathbf{t}^u(s)$ to calculate some desired property of the three-nucleon system, such as the triton energy. We find below that the triton energy is insensitive to the value of the two-body t-matrix at very large values of the momentum. (Lavine et al., 1973; Fuda, 1968B.) Hence the approximation $\mathbf{t}^u$ is "good enough" for the desired purpose; though it may well be unsatisfactory for some other purpose, such as calculation of the triton wave function for large momenta.

We have now found that a (rank-one energy-independent) separable potential has four unusual features. First, this potential becomes more and more non-local at large distances; second, an attractive separable potential cannot have more than one bound state; third, a separable potential acts only in one partial wave; and fourth, the phase shift for a separable potential cannot change sign.

These features may make the reader wonder if a separable approximation would *ever* be useful. It turns out that most of these features are not objectionable for the case of nucleon-nucleon scattering in an S state; though they are certainly hard to overcome for some other problems (Coulomb interaction; nucleon-nucleon scattering in p, d, f etc. states). For the nucleon-nucleon case, the first feature is not serious, because we are fitting the t-matrix, *not* the potential: our assumption (3.26) for the potential turns out to be merely a convenient intermediate step in approximating the t-matrix. Of course this answer that **t** is approximately separable but **v** isn't implies the failure of the Born approximation: for a local potential the Born approximation and a separable approximation are incompatible. The second feature of only one bound state is tailor-made to fit the nucleon-nucleon problem, but obviously gives trouble in many other problems. The third feature is not serious, since nuclear physicists have already given up "strong locality" and use different (weakly local) potentials acting in each partial wave. The fourth feature is certainly serious: we cannot fit the on-shell t-matrix at energies of order two hundred MeV for the nucleon-nucleon case. But *perhaps* this failure will still permit success in fitting off-shell elements of the t-matrix at negative energy or even low positive energy. Mitra (1961, 1962) presented this argument over ten years ago; and by now we have numerical examples that Mitra was right.

3.6 Epistemological Remarks

In an earlier paper (Levinger et al., 1969) we made some methodological and epistemological remarks concerning separable approximations to the off-shell t-matrix. These remarks seem as reasonable now as when they were written four years ago, and are repeated here without major changes. I then discuss a recent approach (Baranger et al., 1969) that emphasizes the half-off shell t-matrix and corresponding eigenfunction $|\psi_k\rangle$.

We first note the phenomenological character of the nucleon-nucleon t-matrix. Experiments (for example) provide $^1S_0$ phase shifts for a finite range of energy. If we know the phase shifts *accurately* at *all* energies, and *if* the potential were known to be local, then Gelfand and Levitan (Goldberger et al., 1964) tell us that this local potential is unique. (However, see Fig. 2.2 for potentials in current use.) We could then use this unique local potential in the partial wave LS equation to find unique values of this partial wave t-matrix. That is, we have used the *assumption* of a "partial wave local" potential to provide a prescription to extrapolate from the on-shell t-matrix (found from the known phase shifts) to determine off-shell values of the t-matrix. If we made another assumption, such as separability of the energy-independent potential and of the t-matrix, we would have another prescription for extrapolation, provided by Bolsterli (1965).

An assumption of locality in a given partial wave should be unnecessary in atomic physics problems, such as scattering of helium atoms by helium atoms, for two reasons. First, we know what two-body potential to insert in the Schrodinger equation. Of course, we face extremely difficult problems in obtaining an exact solution to the helium-helium problem, since it involves 6 particles (4 electrons and 2 α particles). Nevertheless, our ability at least to write down the equation allows us to isolate various terms in the equivalent two-body potential: namely a static term, and a velocity-dependent term due to frequency-dependent polarizability of the two atoms. Second, *in principle* we can measure appreciable phase shifts for a *large* number of different partial waves, while staying at sufficiently low energies that we cannot excite internal degrees of freedom in the atoms. This limitation to energies of order a Rydberg allows us to use roughly $(M_\alpha/m_e)^{\frac{1}{2}} \simeq 85$ partial waves in helium-helium scattering. But the corresponding limitation to center of mass energies of 140 MeV (the pion rest mass) limits us to roughly $(M/m_\pi)^{\frac{1}{2}}$ or $\simeq 3$ different partial waves in the nucleon-nucleon case. Here M_α, M and m_e and m_π refer to masses of α, nucleon, electron and pion respectively. Even this toehold on knowledge of the locality for the nucleon potential is taken away, when we introduce nucleon-nucleon potentials that explicitly depend on the value of the angular momentum: namely, a Majorana exchange potential that is attractive in S states and repulsive in P states, and a quadratic spin-orbit potential that is different in ^1S and ^1D states.

Which prescription for extrapolation of the nucleon-nucleon t-matrix is correct? An epistemological problem cannot be avoided: is there at present *any* method to determine which prescription is the "correct" method of extrapolation of the t-matrix (Levinger, 1973A).

We might try to solve this epistemological problem in one of three ways: Occam's razor, use of other experimental data, or pragmatically.

i) The use of a "partial-wave local" potential does seem the *simplest* assumption thus avoiding needless multiplication of hypotheses. Non-locality is often avoided by the "Pandora's box argument": if I were to admit the possibility of a non-local potential, I would have so many unknown quantities in nuclear physics that I would never be able to calculate anything. So don't open the box (Blatt and Weisskopf, 1952). Another justification for locality has been given in Chapter II: the local one-pion-exchange-potential should be a good approximation to the true nucleon-nucleon potential at reasonably large distances. This *suggests* that the potential might be local for all distances. But the real use of Occam's razor still lies in the future. We must make the simplest assumption of a local potential fitted to the properties of the two-nucleon system, and calculate a variety of other nuclear properties, such as the binding energies, sizes and shapes of specified nuclei. If these calculations can be done accurately, and give good agreement with experimental data, then we are justified in invoking Occam to limit ourselves to the simplest assumption of a local potential. But calculations with a local "realistic" potential have in general not yet been done accurately. Serious approximations are at present necessary, except in the single case of calculation of the ground state of the trinucleon (see Chapter VI). In the trinucleon case, the accurate calculation misses the experimental energy by some $1\frac{1}{2}$ MeV; but this cannot be taken as a strong argument against locality, since there are several possible sources of this discrepancy discussed in Chapter VI.

ii) We might appeal to *other experiments* (e.g., electron-deuteron-scattering) to determine off-shell values of the nucleon-nucleon *t*-matrix. The approach of Chapter II and VII to disentangle the Gordian knot represents a *program* that uses future experiments and certain assumptions in their analysis: it is far from complete.

iii) We might dodge the unsolved epistemological problem in a *pragmatic* (or one might say an opportunistic) manner, by using a separable *t*-matrix as an *approximation* to the "true" *t*-matrix, whatever that mysterious object is. Suppose that *t*-matrices for a local and a separable potential agree "well enough" with each other in a "specified region" of p, k, and s space. Since locality and separability represent radically different assumptions, it would then be *plausible* to assume that all extrapolation methods would agree well enough. We could then argue that the separable *t*-matrix would be "close enough" to the "true" *t*-matrix. But the closer the local and separable approximations agree, the less likely is it that the "true" *t*-matrix lies between them! (Levinger, 1973A.)

I must admit the possibility that we physicists are trying to solve an insoluble problem. There may be no way to disentangle the Gordian knot, in which different extrapolations to find the off-shell two-body *t*-matrix can be compensated for by different choices of the three-nucleon force (Noyes, 1972B).

Also, the extrapolation problem is more complicated than stated above, since we must admit the possibility of energy-dependent non-local two-body potentials.

I am provisionally adopting the third, or pragmatic, viewpoint, while I am exploring the second approach, and leaving the first to many other workers in nuclear theory. In using the third approach, we immediately face the problem of providing meaning for the vague terms used above: "well enough" and "specified region". Of course, an identical problem occurs whenever we use an approximation: e.g., if we assert that classical mechanics works "well enough" in a "specified region" for problems in celestial mechanics. An approximation is always motivated by the desire to solve a certain type of problem: e.g., to find the energy E_T of the trinucleon. So we limit ourselves to negative values of energy s. While we need the t-matrix for all momenta p and k, we can use a poorer approximation for those high values of p and k that occur rarely in the trinucleon ground state wave function (Gupta, 1967). The upper limit chosen for relevant momenta gives us a corresponding lower limit for the range of energy. I *estimate* that the relevant region is

$$0 < p < 3 \, \text{fm}^{-1}$$
$$0 < k < 3 \, \text{fm}^{-1} \tag{3.49}$$
$$-300 < s < -8 \, \text{MeV}$$

or

$$-7 < s < -0.2 \quad \text{in our energy units.}$$

The accuracy needed in extrapolation of the t-matrix clearly need not be an order of magnitude better than our knowledge of the on-shell values. As discussed in Chapter II and V, one main uncertainty in E_T is the choice made of the percentage D-state, p_D: i.e., whether we use a strong central and a weak but long-range tensor force (small p_D), or a weak central and strong tensor force (large p_D). If we choose $p_D = 5\% \pm 2\%$ the estimated uncertainty of 2% produces an uncertainty of 0.5 MeV in E_T (Phillips, 1968; Brady et al., 1969). Uncertainties in knowledge of the singlet effective range, and in the shape of the two-nucleon potentials (see Fig. 2.2) produce uncertainties in E_T of the same order of magnitude. So an approximation that allows us to calculate E_T (for a given choice of p_D, singlet effective range, and potential shape) with an error much less than 0.5 MeV is clearly satisfactory at present. Our extrapolation methods will, hopefully, become more accurate as our knowledge of the on-shell t-matrix improves. We will show below that the separable approximation does give an error much less than 0.5 MeV, at least in the central force case with a soft core: e.g., the Reid singlet.

I shall test the accuracy of our separable approximations by three methods. First, I look at the separable and local values of the off-shell t-matrix for different values of momenta and energy in the relevant range. Second, in the next chapter I use Harms' unitary pole expansion to give a series which we can truncate at any desired point to estimate the error due to truncation. Third, in Chapter V, I compare trinucleon energies calculated with separable and local t-matrices.

An earlier study (Levinger et al., 1969) of the accuracy of the UPA (and of other approximations) for a local central attractive square well turns out not to be of great significance, due to a special feature of this square well noted by Brayshaw et al. (1970): the square well potential is separable, to a good approximation. That is, the S-wave potential in momentum space is separable if we neglect terms of order $(kb)^2$ and $(pb)^2$, where b is the width of the well.

Baranger et al. (1969) have initiated an examination of the arbitrariness of the extrapolation procedure to find the off-shell t-matrix. Their work complements the approach above. I ask, "how accurately does the UPA t-matrix agree with that for a local potential?"; while Baranger asks "what is the possible range of extrapolated values of the off-shell t-matrix, for an energy-independent potential of arbitrary local or non-local character?".

Consider a central force problem with no bound state, such as the ^1S system, with measured phase shifts, which give the on-shell value of the partial wave t-matrix [see Eq. (3.13)]. At a given positive energy k^2, complete knowledge of the quantum mechanical system is provided by the wave function $|\psi_k^+\rangle$; the plus sign shows an outgoing wave. The asymptotic form at large r of $\langle r|\psi_k^+\rangle$ of course is determined by the measured phase shift $\delta(k)$; but the wave function itself cannot be determined in a two-body experiment. This unknown wave function determines the half-off-shell t-matrix through the relation

$$\langle p|\psi_k^+\rangle = \langle p|k\rangle + \langle k^2 - p^2 + i\varepsilon\rangle^{-1}\, t(p, k; k^2 + i\varepsilon)\,. \tag{3.50}$$

The complex half-off-shell t-matrix can be written as a product of a real function $\phi(k, p)$ and the complex number $\exp[i\delta(k)]$. The completely off-shell t-matrix can then be found from half-off-shell values from the dispersion relation

$$t(p, k; s) = \phi(k, p)\cos\delta(k)$$
$$+ \int_0^\infty dq[(s - q^2)^{-1} - P(k^2 - q^2)^{-1}]\,\phi(q, p)\,\phi(q, k)\,. \tag{3.51}$$

Here P denotes the Cauchy principle value.

The function $\phi(k, p)$ is expressed as the sum of a term symmetric in interchange of k and p, and an antisymmetric term. The symmetric term is arbitrary, except for the values at $k = p$ known from the phase shifts. The antisymmetric term can be determined from the symmetric term.

This approach is very helpful in understanding the arbitrariness of the off-shell t-matrix: we do not know the wavefunction $|\psi_k^+\rangle$, except asymptotically at large r. The arbitrary behavior at small r [or large momentum p in (3.50)] gives an arbitrary behavior in the symmetric part of the function $\phi(p, k)$ and of the completely off-shell t-matrix $t(p, k; s)$. [Thus we can focus on our understanding of elementary quantum mechanics, which treats wave functions, rather than t-matrices. This emphasis on the wave function also provides a simple method, in principle, of determining whether the potential is indeed energy-independent, as assumed above. Namely, measure the wave functions for two different energies, and determine whether they obey the appropriate orthogonality relation for a (local or non-local) energy-independent potential. The orthogonality relation is particularly simple for two bound states at different energies: cf. the argument above showing that an energy-independent separable potential could have only one bound state.]

I shall not attempt here to summarize the extensive literature following Baranger's paper. (For instance, Amado, 1970; van Dijk et al., 1970; Kowalski, 1971; Haftel, 1970; Haftel, 1973.) I mention three recent papers that provide numerical examples of the extrapolation procedure. Picker et al. (1971) introduce arbitrary parameters into the wave function $|\psi_k^+\rangle$. On the other hand Sauer (1973) and Sauer et al. (1973) introduce adjustable parameters in the symmetric part of the function $\phi(p, k)$, and calculate the effect of this arbitrariness on the triton energy.

One problem that is not yet completely solved is how to introduce the constraint that the potential is presumed to be local, and of value given by OPEP, at moderately large r. That is, we know more about $\langle r | \psi_k^+ \rangle$ than merely its asymptotic value at very large r. See Sauer (1973).

If the two-body system has a bound state, Haftel's (1971, 1973) technique of generating phase-equivalent potentials by the use of a short-range unitary transformation is particularly useful. (See Eqs. 2.9 to 2.11, and Figs. 2.4 and 2.13.)

3.7 The Unitary Pole Approximation (UPA)

We noted above that delta function potential was exactly separable. Suppose, on the other hand, that we are given a non-separable potential (which may well be local, but need not be) that has a single bound state at energy $-B$, with wave function $|B\rangle$, and a set of S-wave phase shifts $\delta(k)$, for real k. We might use either the bound state $|B\rangle$ or the phase shifts $\delta(k)$ to find schemes to approximate the exact t-matrix $t(p, k; s)$ by a rank-one separable expression. The UPA prescription fits the exact t-matrix at its pole $s = -B$. A second scheme, using the phase shift, separates into two different prescriptions, depending on whether we demand that the separable potential be energy independent. If we insist on energy independence of the form factor $g(p)$, then we have the Bolsterli prescription for deriving a separable potential from a set of phase shifts (Bolsterli et al., 1965). If we allow energy dependence of the form factor $g(p, s)$ then we have the Noyes scheme (Noyes, 1965; Kowalski, 1965). I concentrate below on the UPA.

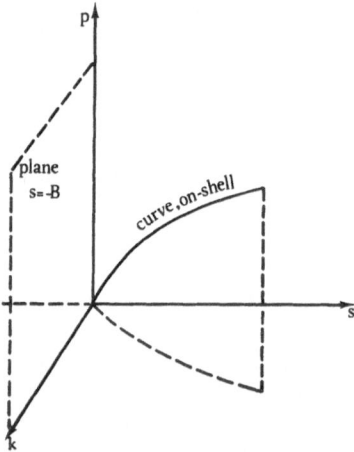

Fig. 3.1. Sketch of p, k, s space; where p and k are momenta and s is the energy used in the t-matrix. The curve shows on-shell, $p^2 = k^2 = s$. The pk plane at the energy of the bound state pole shows the region where the unitary pole approximation is accurate. From Levinger (1973 A)

Fitting the t-matrix at its pole, gives us a separable t_u that fits $t(p, k; s)$ exactly at *all* values of momenta p and k, at energy $s = -B$. The pole fit is then exact in the $p - k$ *plane* designated as UPA in Fig. 3.1. On the other hand, a fit to on-shell values, means an exact fit on the *curve* marked OS in the same figure. This drawing *suggests* that for an approximation to be used in the region of negative s, the pole fit would be more desirable than the on-shell fit (Levinger, 1973A). Of course this suggestion should be tested either by general mathematical arguments, or by an examination of specific examples.

The pole fit (or UPA) due to Lovelace (1964) and Fuda (1968A) merely consists of using Yamaguchi's equations [Eqs. (3.30) and (3.31)] the other way around. Earlier we assumed we had a separable potential (3.26) with strength λ and form factor $g(p)$. This potential substituted into the momentum-space Schrödinger equation gave us the energy eigenvalue $E = -\gamma^2$ and the momentum-space wave function $\phi(p)$ in terms of λ and $g(p)$. For the present purpose, we regard γ and $\phi(p)$ as given from the solution for the assumed non-separable potential, and solve (3.30) to find the form factor (within a proportionality constant):

$$g(p) = (\gamma^2 + p^2)\,\phi(p)\,. \tag{3.52}$$

Given γ and $g(p)$ we use Eq. (3.31) to find the strength $\lambda(\gamma)$. We then use our determined form factor and strength to find the separable UPA t-matrix $t_U(p, k; s) = -g(p)\,g(k)/D(s)$ where the energy dependent denominator $D(s)$ is given in (3.40).

We are approximating the t-matrix for a local potential at the energy where it has a pole, and is therefore separable, from Eq. (3.15). Why do we use the term "unitary"? Our t-matrix $\mathbf{t}_U$ is unitary since it is the solution of the Schrödinger equation for a Hermitian (but non-local) potential. Note that we would have a poor approximation if we used the energy dependence $(s + B)^{-1}$ of Eq. (3.15); we have a much better approximation using $1/D(s)$, from (3.40).

Harms (1969, 1970) has studied several local potentials as tests of the accuracy of the UPA. The Hulthén potential is of special interest, since substitution of its momentum space wave function for the bound state in (3.52) gives just the Yamaguchi (1954) form factor, $g(p) = 1/(p^2 + \beta^2)$. This particular form factor has been used extensively in two and three-nucleon work. It has an advantage in that the integral for $D(s)$ can be performed analytically, and so can some of the needed integrals in the three-nucleon problem.

I present below numerical examples of the accuracy of the UPA for a Yukawa potential, and for the Reid soft core singlet potential (which is a combination of several Yukawa potentials). Substituting a Yukawa local potential $v(r) = -V_0 \exp(-\mu r)/\mu r$, (3.12) gives the S-wave potential in momentum space as

$$v(p, k) = -(V_0/2\pi\mu pk)\ln\{[\mu^2 + (p + k)^2]/[\mu^2 + (p - k)^2]\}\,. \tag{3.53}$$

Harms chooses $\mu = 0.633\,\mathrm{fm}^{-1}$, and $V_0 = 0.9950$, giving $B = 2.225\,\mathrm{MeV}$. Values for $v(p, k)$ are given in Table 3.1 and Figs. 3.2 to 3.5. We compare the potential with the t-matrix $t(p, k; s)$ and with the UPA approximation to the t-matrix, $t_U(p, k; s)$.

Harms (1969) solves the LS equation (3.11) for the t-matrix for a specified negative energy s by standard numerical techniques. He uses a quadrature formula to replace the integral by a finite sum; the integral equation then becomes an equation in which the unknown matrix $\mathbf{t}(s)$ is expressed in terms of known matrices $\mathbf{v}$ and $\mathbf{G}_0(s)$. This matrix

Table 3.1. UPA for Yukawa potential

p	k	$v(p,k)$	$t_U(p,k;-0.1)$	$t(p,k;-0.1)$	$t_U(p,k;-0.5)$	$t(p,k;-0.5)$	$t_U(p,k;-3)$	$t(p,k;-3)$
0.003	0.003	−2.49799	−14.0242	−14.4024	−3.4755	−4.0667	−1.9775	−2.7962
0.104	0.003	−2.43296	−13.9119	−14.2520	−3.4477	−3.9934	−1.9617	−2.7298
0.104	0.104	−2.37278	−13.8006	−14.1077	−3.4201	−3.9256	−1.9460	−2.6688
0.385	0.003	−1.82394	−12.6441	−12.6689	−3.1335	−3.2914	−1.7829	−2.1089
0.385	0.104	−1.80107	−12.5429	−12.5697	−3.1084	−3.2616	−1.7686	−2.0852
0.385	0.385	−1.53352	−11.3999	−11.4255	−2.8252	−2.9111	−1.6075	−1.8069
0.953	0.003	−0.76491	−8.5849	−8.3264	−2.1275	−1.8734	−1.2105	−0.9980
0.953	0.104	−0.76443	−8.5161	−8.2747	−2.1105	−1.8685	−1.2008	−0.9970
0.953	0.385	−0.75603	−7.7401	−7.6764	−1.9182	−1.8073	−1.0914	−0.9824
0.953	0.953	−0.63641	−5.2552	−5.5332	−1.3024	−1.4678	−0.7410	−0.8328
2.092	0.003	−0.20958	−3.6623	−3.5188	−0.9076	−0.7455	−0.5164	−0.3414
2.092	0.104	−0.20969	−3.6330	−3.4973	−0.9003	−0.7438	−0.5123	−0.3413
2.092	0.385	−0.21102	−3.3019	−3.2492	−0.8183	−0.7242	−0.4656	−0.3407
2.092	0.953	−0.21843	−2.2419	−2.3986	−0.5556	−0.6430	−0.3161	−0.3381
2.092	2.092	−0.21732	−0.9564	−1.2243	−0.2370	−0.4614	−0.1349	−0.3039
4.665	0.003	−0.04517	−0.9751	−0.9368	−0.2417	−0.1979	−0.1375	−0.0885
4.665	0.104	−0.04518	−0.9673	−0.9310	−0.2397	−0.1975	−0.1364	−0.0885
4.665	0.385	−0.04527	−0.8792	−0.8648	−0.2179	−0.1920	−0.1240	−0.0881
4.665	0.953	−0.04576	−0.5969	−0.6372	−0.1479	−0.1692	−0.0842	−0.0861
4.665	2.092	−0.04824	−0.2546	−0.3287	−0.0631	−0.1251	−0.0359	−0.0809
4.665	4.665	−0.06194	−0.0678	−0.1476	−0.0168	−0.0930	−0.0096	−0.0797

Momenta p and k are given in fm^{-1}. The S-wave Yukawa potential in momentum space $v(p,k)$ is given in fm^{-1}; see Eq. (3.53). $t_U(p,k;s)$ is the off-shell t-matrix in fm^{-2} in the unitary pole approximation, where s is the energy in fm^{-2}. $t(p,k;s)$ is the t-matrix found by numerical solution of the Lippmann-Schwinger equation. From Harms (1970).

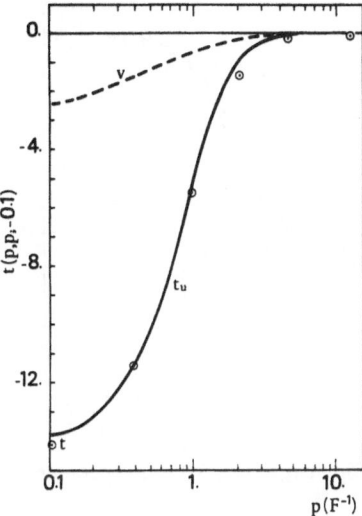

Fig. 3.2. Comparison of the momentum space potential (dashed curve), the unitary pole approximation to the t-matrix (solid curve) and numerical values of t (circled points), all for a Yukawa potential. The momenta p and k are equal; the energy $s = -0.1$. From Harms (1970)

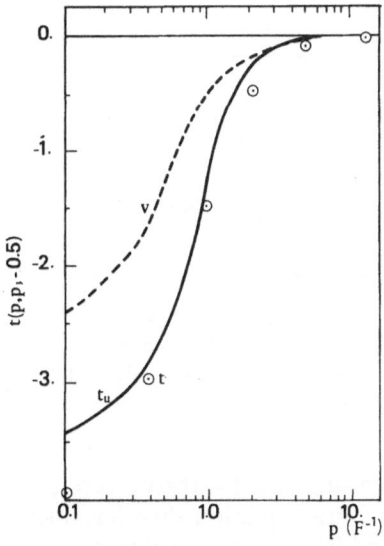

Fig. 3.3. Yukawa potential, for energy $s = -0.5$; same notation as Fig. 3.2. From Harms (1970)

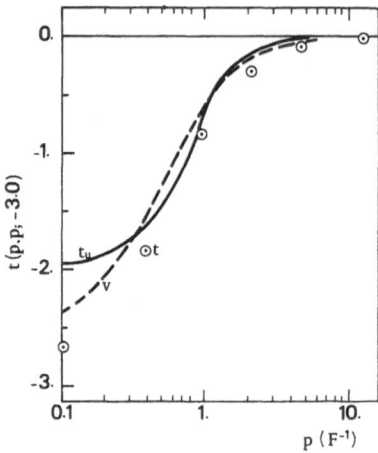

Fig. 3.4. Yukawa potential, for energy $s = -3.0$; same notation as Fig. 3.2. From Harms (1970)

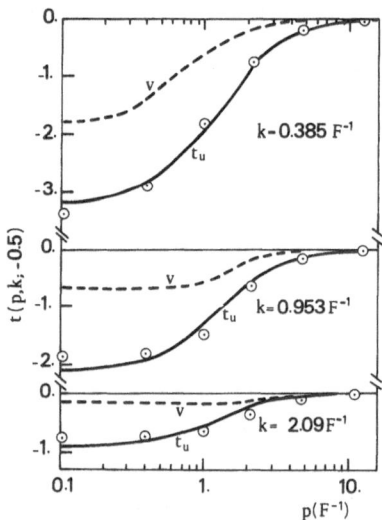

Fig. 3.5. Comparison of off-diagonal elements of t-matrices and potential, for $k = 0.385$, 0.953 and 2.09 fm^{-1}; all for energy $s = -0.5$ for Yukawa potential. For notation see Fig. 3.2. From Harms (1970)

equation is easily solved numerically. (More care is needed for positive values of s, since the integral equation then has a singularity at $q = s^{\frac{1}{2}}$.) We show in the table and figure Harms' values for the t-matrix. Note the large differences between $\mathbf{t}(s)$ and $\mathbf{v}$, even for $s = -3$, corresponding to a CMS energy of -134 MeV: i.e., the Born approximation is poor.

Harms solves the momentum space Schrödinger equation numerically to find the momentum space wave function, and the corresponding form factor $g(p)$. He also integrates (3.40) numerically to obtain the denominator function $D(s)$, and chooses the strength λ of the separable potential so that $D(-B) = 0$. This numerical work gives the UPA t-matrix, also given in Table 3.1 and the figures. The "UPA potential" $-\lambda g(p)\, g(k)$ seriously disagrees with the local potential $v(p, k)$.

It is difficult to compare graphically two functions $\mathbf{t}$ and $\mathbf{t}_U$ each a function of three variables. We shall follow Harms (1969) in showing a few examples. The reader can easily use the tables to make more comparisons on his own. Figure 3.2 shows $t(p, p; -0.1)$ as circled points, $t_U(p, p; -0.1)$ as a solid curve, and $v(p, k)$ as a dashed curve. Since we are near the pole ($s = -B = -0.053$ in our units) the Born approximation is terrible, and the UPA is excellent. The UPA is less successful at $s = -0.5$, and still worse at $s = -3.0$, as shown in Figs. 3.3 and 3.4. One might wonder if the "diagonal values" $p = k$ represent a special case. In Fig. 3.5 we look at several cases of $t(p, k; -0.5)$ for $k = 0.385$, 0.953, and 2.09 fm^{-1} respectively. Again the UPA works "moderately well".

Harms (1970) and Harms et al. (1970) use the methods described above to treat the Reid singlet potential which is written in our units as

$$V(r) = -0.252 \exp(-x)/x - 159 \exp(-4x)/4x + 1095 \exp(-7x)/7x ,$$
$$x = 0.70r . \tag{3.54}$$

The potential in momentum space is just the sum of three terms, of form 3.53. It is positive everywhere, due to the large repulsive term at the right: i.e., the "soft core" is really rather hard. The exact t-matrix $t(p, k; s)$ is found using standard numerical methods. Since this potential is not quite strong enough to give a bound state, Harms proceeds to solve a slightly different problem. He considers a potential of the same shape as 3.54, but stronger by a factor 1.0819, so it will give a bound state at $s = 0$. He then finds the momentum space wave function, and corresponding form factor, for this state at energy zero. The strength λ of the separable potential is then decreased to give the correct position of the anti-bound state. Harms et al. (1970) give a nine term expression for the form factor $g(p)$:

$$g(p) = \sum_{n=1}^{9} C_n/(p^2 + \mu_n^2) . \tag{3.55}$$

The eighteen parameters C_n and μ_n are given in Table 3.2. This form factor is plotted in Fig. 3.6, where it is compared with the Yamaguchi (1954) form factor for the same value of the effective range. Note that

Table 3.2. Parameters for an analytic approximation to the form factor for the Reid singlet potential

n	$\mu_n(fm^{-1})$	C_n
1	0.7	5.6155 − 3
2	1.4	2.8853 0
3	2.1	− 2.3494 + 1
4	2.8	8.5698 + 1
5	5.6	− 5.3856 + 2
6	8.4	2.1009 + 3
7	4.9	− 1.3805 + 1
8	9.8	− 1.8516 + 3
9	14.7	2.4047 + 2

These parameters are used in Eq. (3.55) to give the form factor. See Fig. 3.6.

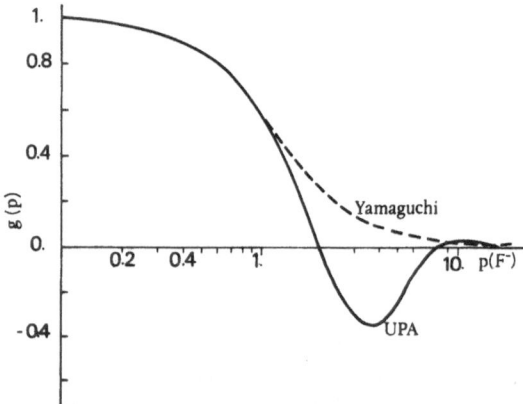

Fig. 3.6. Form factors for unitary pole approximation (solid curve) to Reid singlet soft core potential, and Yamaguchi form factor (dotted curve). From Harms et al. (1970)

the Harms form factor has a node at p near $2\,\mathrm{fm}^{-1}$, while the Yamaguchi form factor has no nodes. This node in $g(p)$ gives a node in the UPA phase shift shown in Fig. 3.7, at a laboratory energy near the 260 MeV where the Reid and experimental phase shifts change sign. As discussed above, the UPA phase shift cannot change sign; but the fit to the Reid phase shifts is reasonable – though not excellent – and represents a great improvement over the use of a Yamaguchi form factor.

Harms results (1970) for the UPA t-matrix at negative s are compared with the exact $t(p, k; s)$ and with the Reid singlet soft core potential $v(p, k)$ in Table 3.3. Note that the "UPA potential" is negative for $p < 2\,\mathrm{fm}^{-1}$ and $k < 2\,\mathrm{fm}^{-1}$ even disagreeing in sign with the positive

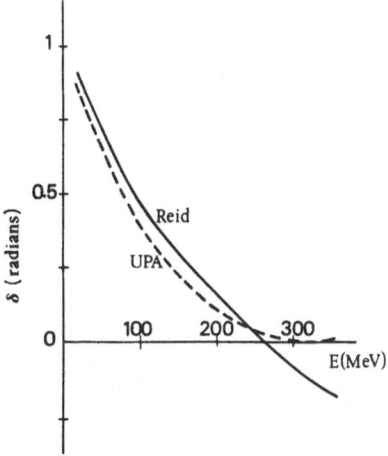

Fig. 3.7. Comparison of the singlet-S phase shifts for the Reid local soft-core potential (solid curve) and by its unitary pole approximation (dashed curve). From Harms et al. (1970)

potential $v(p, k)$. Figures 3.8 to 3.11 illustrate this comparison for $p = k$, for values of energy $s = 0.0$, -0.1, -0.5, and -3.0 respectively. The potential v is not shown since it is off the paper: the Born approximation is impossibly bad! Figure 3.12 shows some off-diagonal values $(p \neq k)$ for $s = -0.5$. The agreement is excellent, for momenta values $p < 2\,\mathrm{fm}^{-1}$ and $k < 2\,\mathrm{fm}^{-1}$: i.e., for the same range of moderate success in fitting the on-shell elements $t(p, p; p^2)$ shown in Fig. 3.7. We face the same sort of problem at negative energy for $t_U(p, p; s)$: namely that t_U is always negative, while the exact solution does change sign at p about $2\,\mathrm{fm}^{-1}$. This problem does not arise in $t(p, 0.567; -0.5)$, as illustrated in the top of Fig. 3.12. The change of sign of $\mathbf{t}_U$ from the factor $g(p)$ matches well with the sign change of t. Agreement between $\mathbf{t}_U$ and $\mathbf{t}$ is not as good at the same energy for $k = 1.41\,\mathrm{fm}^{-1}$, and is still worse for $k = 3.08\,\mathrm{fm}^{-1}$.

Looking at these many figures and tables once more, one finds that for modest values of p and k (below $2\,\mathrm{fm}^{-1}$) the UPA fit to the Reid t-matrix at negative energy is better than the UPA fit to the Yukawa t-matrix. (At higher momenta the situation is reversed.) Since these modest values of momenta dominate in the calculation of the trinucleon ground state [Eq. (3.49)], we might anticipate that the UPA will work fairly well for the Yukawa case, and much better for the Reid singlet case. See Chapter V for the accuracy of the UPA in calculations of E_T.

On the other hand, inaccuracies of the UPA would be expected to appear in calculations where high values of p or k were needed: e.g., electron-trinucleon elastic scattering with high momentum transfer,

Table 3.3. v, t_U and t for Reid potential

p	k	$v(p,k)$	$t_U(p,k;0)$	$t(p,k;0)$	$t_U(p,k;-0.1)$	$t(p,k;-0.1)$	$t_U(p,k;-0.5)$	$t(p,k;-0.5)$	$t_U(p,k;-3.0)$	$t(p,k;-3.0)$
0.004	0.004	0.81769	-10.9533	-10.9529	-2.2704	-2.3002	-1.5391	-1.5908	-1.1588	-1.2341
0.153	0.004	0.86926	-10.7799	-10.7642	-2.2345	-2.2558	-1.5147	-1.5561	-1.1404	-1.2031
0.153	0.153	0.89535	-10.6092	-10.5812	-2.1991	-2.2147	-1.4908	-1.5244	-1.1224	-1.1752
0.567	0.004	1.12867	-8.9424	-8.8170	-1.8536	-1.8215	-1.2566	-1.2311	-0.9460	-0.9224
0.567	0.153	1.13907	-8.8008	-8.6774	-1.8242	-1.7985	-1.2367	-1.2161	-0.9311	-0.9105
0.567	0.567	1.26654	-7.3006	-7.1808	-1.5133	-1.5244	-1.0259	-1.0324	-0.7724	-0.7643
1.405	0.004	1.70973	-3.1890	-2.9760	-0.6610	-0.6076	-0.4481	-0.3985	-0.3374	-0.2626
1.405	0.153	1.71159	-3.1385	-2.9299	-0.6506	-0.6009	-0.4410	-0.3945	-0.3320	-0.2597
1.405	0.567	1.73413	-2.6036	-2.4363	-0.5397	-0.5208	-0.3658	-0.3457	-0.2754	-0.2243
1.405	1.405	1.85721	-0.9285	0.8258	-0.1925	-0.1764	-0.1305	-0.1112	-0.0982	-0.0433
3.084	0.004	2.13370	3.7302	3.8626	0.7732	0.7836	0.5242	0.5179	0.3946	0.4166
3.084	0.153	2.13258	3.6712	3.8050	0.7610	0.7773	0.5159	0.5152	0.3884	0.4154
3.084	0.567	2.11880	3.0454	3.1927	0.6313	0.7031	0.4279	0.4825	0.3222	0.4008
3.084	1.405	2.04551	1.0861	1.2548	0.2251	0.4130	0.1526	0.3400	0.1149	0.3318
3.084	3.084	1.84224	-1.2704	-1.1224	-0.2633	-0.0223	-0.1785	0.0921	-0.1344	0.1903
6.878	0.004	1.33278	0.6971	0.7706	0.1445	0.1560	0.0979	0.1038	0.0737	0.0921
6.878	0.153	1.33269	0.6860	0.7593	0.1422	0.1550	0.0964	0.1035	0.0726	0.0921
6.878	0.567	1.33120	0.5691	0.6398	0.1180	0.1429	0.0800	0.0998	0.0602	0.0921
6.878	1.405	1.32223	0.2029	0.2664	0.0421	0.0987	0.0285	0.0854	0.0215	0.0926
6.878	3.084	1.27858	-0.2374	-0.1750	-0.0492	0.0452	-0.0334	0.0699	-0.0251	0.0988
6.878	6.878	1.01729	-0.0444	0.0294	-0.0092	0.0736	-0.0062	0.0794	-0.0047	0.0902

Momenta p and k are given in fm^{-1}. The potential in momentum space $v(p,k)$ for the Reid soft core ^1S potential is found from Eqs. (3.53) and (3.54). $t_U(p,k;s)$ and $t(p,k;s)$ are the unitary pole approximation and numerical values of the t-matrix at energy s.

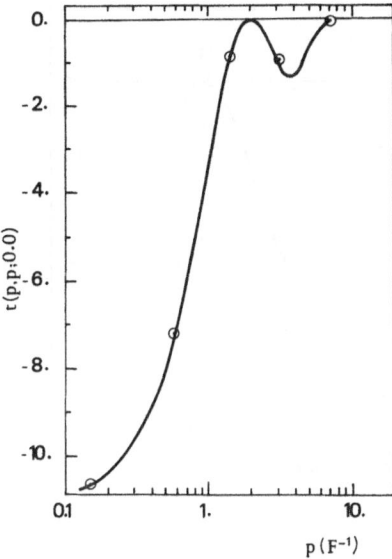

Fig. 3.8. Comparison of the unitary pole approximation (solid curve) and numerical values (circled points) for the Reid singlet soft core potential. We choose equal momenta p and k; the energy $s = 0$. From Harms (1970)

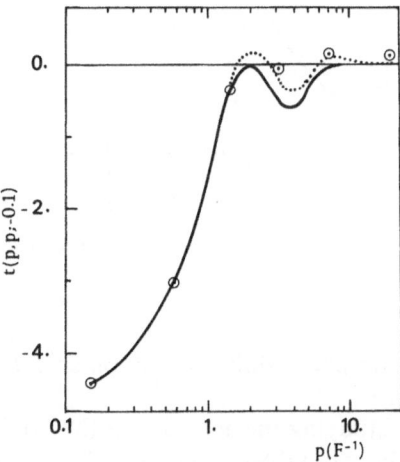

Fig. 3.9. Comparison of unitary pole approximation (solid curve) and numerical values of the t-matrix for the Reid singlet soft core potential (circled points) for equal momenta, and for energy $s = -0.1$. The dotted curve shows three terms in the unitary pole expansion (Chapter IV). From Harms (1970)

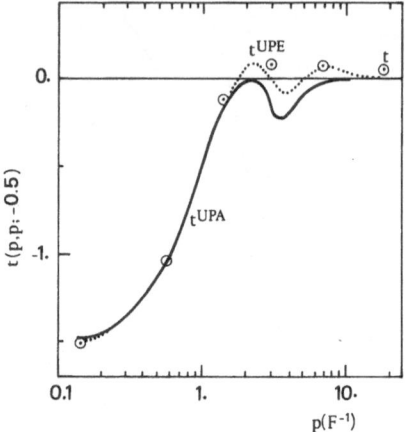

Fig. 3.10. Reid soft core singlet; notation same as Fig. 3.9 for energy $s = -0.5$. From Harms (1970)

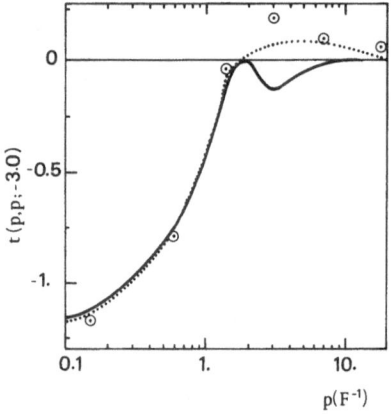

Fig. 3.11. Reid soft core singlet; notation same as Fig. 3.9 for energy $s = -3.0$. From Harms (1970)

high energy two-nucleon phase shifts, or half-off shell t-matrices at high values of energy.

A main trouble in studying the accuracy of the UPA is that we tend to get buried by numbers of tables or graphs. The quantity of numbers to compare is even worse at positive energy; since now we should in general compare two *complex* functions t_U and t. We have avoided this difficulty for the on-shell values, by using the real phase shifts shown in Fig. 3.7. We can also avoid it for half-off shell values of the t-matrix by

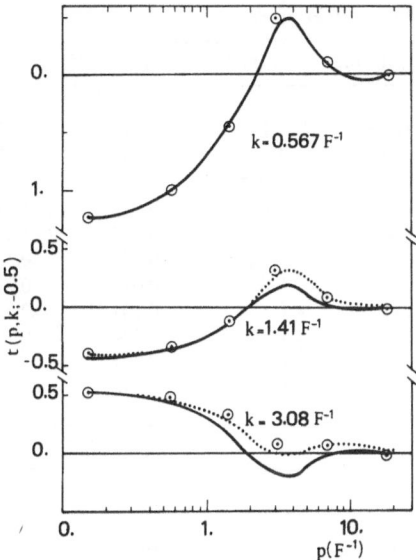

Fig. 3.12. Reid soft core singlet; off-diagonal elements of the t-matrices, for energy $s = -0.5$; see Fig. 3.9 for notation. From Harms (1970)

comparison of the real Noyes (1965) half off-shell function, or Baranger's function $\phi(p, k)$. The former is defined as

$$f(p, k) \equiv t(p, k; k^2)/t(k, k; k^2) . \tag{3.56}$$

Numerical values of the on-shell and half-off-shell t-matrices for the Reid singlet potential were provided by Redish (private communication); while the UPA values are found trivially from (3.39):

$$f_U(p, k) = g(p)/g(k) . \tag{3.57}$$

We (Levinger et al., 1972) take the form factor $g(p)$ from (3.55). Table 3.4 compares the values of f and f_U; some results are illustrated in Fig. 3.13.

We see that for $k = 0.3\,\text{fm}^{-1}$, the UPA is quite accurate. The UPA becomes less satisfactory at higher values of k, but is qualitatively useful even at $k = 1.5\,\text{fm}^{-1}$. The UPA is terrible beyond $k = 2\,\text{fm}^{-1}$.

We have emphasized the UPA which fits the bound state pole, to obtain a t-matrix which we hope will be good at negative energies. We should at least acknowledge the existence of other schemes, which instead emphasize fitting the t-matrix at positive energies, with a separable form. While these "positive energy fits" should work better than the UPA in calculations utilizing the t-matrix at positive energy, they may be less successful than the UPA in calculations of the trinucleon

Table 3.4. Accuracy of UPA for Reid potential

$p(F^{-1})$	$f(p, 0.3)$	$f_U(p, 0.3)$	$f(p, 0.75)$	$f_U(p, 0.75)$	$f(p, 1.2)$	$f_U(p, 1.2)$
0.0	1.053	1.059	1.256	1.412	1.856	2.307
0.3	1.000	1.000	1.217	1.333	1.801	2.178
0.6	0.856	0.851	1.095	1.134	1.64	1.854
0.9	0.658	0.638	0.885	0.851	1.375	1.390
1.2	0.444	0.459	0.614	0.612	1.000	1.000
1.5	0.236	0.246	0.327	0.328	0.539	0.537
1.8	0.047	0.054	0.059	0.072	0.055	0.117
2.1	-0.112	-0.081	-0.174	-0.107	-0.390	-0.176
2.4	-0.237	-0.206	-0.364	-0.274	-0.763	-0.449
2.7	-0.327	-0.296	-0.500	-0.394	-1.051	-0.644
3.0	-0.386	-0.352	-0.591	-0.470	-1.249	-0.768

$p(F^{-1})$	$f(p, 1.5)$	$f_U(p, 1.5)$	$f(p, 2.1)$	$f_U(p, 2.1)$	$f(p, 2.7)$	$f_U(p, 2.7)$
0.0	3.55	4.30	-0.466	-13.14	0.622	-3.58
0.3	3.45	4.06	-0.442	-12.4	0.621	-3.38
0.6	3.14	3.45	-0.365	-10.6	0.619	-2.88
0.9	2.63	2.59	-0.232	-7.92	0.620	-2.16
1.2	1.92	1.86	-0.033	-5.69	0.630	-1.55
1.5	1.000	1.000	0.237	-3.06	0.656	-0.833
1.8	-0.067	0.22	0.581	-0.67	0.704	-0.182
2.1	-1.14	-0.33	1.000	1.000	0.777	0.272
2.4	-2.08	-0.84	1.458	2.56	0.877	0.697
2.7	-2.84	-1.20	1.88	3.67	1.000	1.000
3.0	-3.38	-1.43	2.22	4.38	1.13	1.19

$f(p, k)$ is the half-off-shell Noyes function for momenta p and k; f_U is the Noyes function using the unitary pole approximation. See Eqs. (3.56) and (3.57).

energy E_T. Consider a potential which gives a phase shift that does not change sign. We then have two different prescriptions for finding a separable t-matrix that reproduces these phase shifts exactly: First, we can use Bolsterli's scheme (1965), in which the form factor for the separable potential is expressed as an integral involving the phase shift. Second, we can use the Noyes-Kowalski scheme (1965) in which it is assumed that the half-off shell function (3.56) is separable.

Bolsterli and Mackenzie (1965) give an explicit solution to the problem inverse to that already solved in Eq. 3.45. Instead of finding the phase shift given the form factor $g(k)$ and the strength λ of a (rank-one energy independent central) separable potential, they suppose the phase shift $\delta(k)$ is given at all energies, and ask for the separable potential that will exactly produce this phase shift. Of course, the phase shift should not change sign. Consider the simpler problem in which there is no

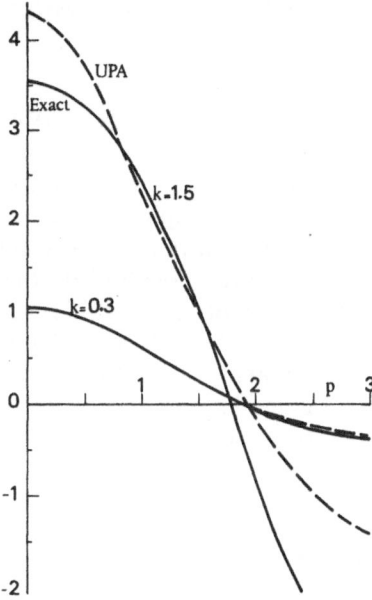

Fig. 3.13. Comparison of Noyes half-off shell functions for the Reid soft core potential. The solid curves show the exact values for on-shell momentum $k = 0.3$ and $1.5 \, \mathrm{fm}^{-1}$ respectively; the dashed curves show the unitary pole approximation. From Levinger et al. (1972)

bound state. Bolsterli et al. find (in our notation)

$$\lambda g^2(k) = (4\pi^2) \, k^{-1} \sin \delta(k) \exp(P \tilde{\delta}(k^2)) \tag{3.58}$$

where

$$\tilde{\delta}(k^2) = \pi^{-1} \int_0^\infty \delta(E')/(k^2 - E') \, \mathrm{d}E' . \tag{3.59}$$

The condition that the form factor $g(k)$ be real imposes the condition that the phase shift cannot change sign.

If the phase shift *does* change sign (as in the ^1S case) we can still fit the on-shell values of the t-matrix with a rank-one separable potential, provided that it is energy-dependent. The simplest scheme is to allow the strength $\lambda(s)$ to depend on s: e.g., it could change sign where the phase shift changes sign (BolM 65), or λ could be a continuous function of s. Alternatively, the form factor $g(k, s)$ might vary with energy, as in the scheme of Noyes (1965) or Kowalski (1965) recently extended by Bhatia et al. (1972).

Alternatively, we could fit a phase shift with a node with an energy-independent non-local potential, provided that we are willing to use a rank-2 separable potential. I pursue this possibility in the next chapter.

IV. More Complicated Separable Potentials

4.1 Introduction

In the previous chapter we developed the UPA for a central local potential with a bound state with wave function $|B\rangle$; we found that the approximation was quite good for the t-matrix at negative energy. Here I shall discuss two more complicated problems: the UPA for a tensor local potential; and use of a rank-N separable potential to give a better approximation [the Harms' (1970) UPE, the Fiedeldey (1969) — Fuda (1970) approximation or the Ernst (1973) approximation] to the t-matrix for a central potential. Rank-N separable potentials can also be used for tensor forces (Tabakin, 1964; Mongan, 1968 and 1969) and Mitra (1959) showed long ago how to use a form of separable potential for a linear spin-orbit force. But I shall not treat these problems in detail.

4.2 UPA for Non-Central Forces

I already noted that the Low equation showed that the t-matrix would be approximately separable near a pole, with the form given by (3.16) (Fuda, 1968 A). Matching the form factor $g(p)$ to (3.16) we have a simple generalization of (3.17), namely

$$g(p) = \langle p|V|B\rangle = -(p^2 + B)\langle p|B\rangle$$
$$= -(p^2 + B)\,\phi(p)\,. \tag{4.1}$$

Here the momentum space wave function for a non-central potential has the form

$$\phi(p) = N[\phi_0(p) + S_{12}(\hat{p})\,\phi_2(p)/\sqrt{8}] \tag{4.2}$$

where N is a normalization factor, $S_{12}(\hat{p})$ is the tensor operator that depends on the unit vector $\hat{p}$, and $\phi_0(p)$ and $\phi_2(p)$ are the S-wave and D-wave functions in momentum space. They are related to the more familiar wave functions $u(r)$ and $w(r)$ by Fourier-Bessel transforms. Define $\psi_0(r) = u(r)/r$ and $\psi_2(r) = w(r)/r$. Then for $l = 0$ or 2,

$$\psi_l(r) = (2/\pi)^{\frac{1}{2}} \int_0^\infty \phi_l(p)\, i^l j_l(pr)\, p^2\, \mathrm{d}p\,. \tag{4.3}$$

We follow the Yamaguchi's (1954) in equating (4.1) and (4.2). We see that the form factor $g(p)$ should be expressed in the same form as (4.2).

$$g(p) = g_0(p) + S_{12}(\hat{p})\, g_2(p)/\sqrt{8}\,. \tag{4.4}$$

We equate S-wave and D-wave dependence separately, giving

$$g_l(p) = -(\gamma^2 + p^2)\, \phi_l(p) \tag{4.5}$$

in close analogy with (3.17).

I define a separable (rank-one, energy-independent) tensor potential as one that has the form

$$v(\mathbf{p}, \mathbf{k}) = -\lambda g(\mathbf{p})\, g(\mathbf{k})\,. \tag{4.6}$$

This can be substituted into the LS equation for tensor forces to give a separable t-matrix. Since we are now dealing with a non-central force, we must find matrix elements of both the potential v and the t-matrix between states that can have different values of angular momentum L: i.e., between L and L', where for 1^+ states L and L' can each take the value 0 or 2.

For the separable potential (4.6),

$$v_{LL'}(p, k) = -\lambda g_L(p)\, g_{L'}(k)\,. \tag{4.7}$$

The LS equation (3.8), when sandwiched between bra $\langle p, L|$ and ket $|k, L'\rangle$ reads

$$\begin{aligned}
\langle p, L|t(s)|k, L'\rangle &\equiv t_{LL'}(p, k; s) = v_{LL'}(p, k) \\
&+ \sum_l \int_0^\infty dq\, q^2\, v_{Ll}(p, q)\, t_{lL'}(q, k; s)\,(s - q^2)^{-1}\,.
\end{aligned} \tag{4.8}$$

Here the summation over angular momentum l includes the two cases, 0 and 2.

We substitute the separable form (4.7) into the LS equation, and obtain the result

$$t_{LL'}^U(p, k; s) = -g_L(p)\, g_{L'}(k)/D_T(s)\,. \tag{4.9}$$

The numerator looks just like our earlier result (3.39) for the central separable case. Our present denominator is a bit more complicated. Here

$$D_T(s) = \lambda^{-1} + \sum_l \int_0^\infty dq\, q^2 g_l^2(q)/(s - q^2) \tag{4.10}$$

where the sum includes the cases 0 and 2.

The Yamaguchi's (1954) chose the same $g_0 = 1/(p^2 + \beta^2)$ discussed in Chapter III; they choose $g_2(p) = -tp^2(p^2 + v^2)^{-2}$. Then the integrals in (4.9) can be done analytically (Siebert, 1973). Phillips (1968) and Brady et al. (1969) varied the numerical values of .parameters, to fit the deuteron static properties with a range of values of p_D.

We now have all the mathematics to explain Burnap's (1970) and Brady's (1972) work with Yamaguchi wave functions, discussed in

Chapter II. Namely, Brady uses (4.5) and (4.3) to find wave functions $u(r)$ and $w(r)$; he then uses these wave functions to find the monopole and quadrupole form factors, and then the extrapolated deuteron tensor polarization. Burnap substitutes the wave function $u(r)$ and $w(r)$ into the Schrödinger equation for non-central forces to find the local central and tensor potentials, to compare with OPEP.

We now develop the UPA for a tensor force for a local non-central potential. As discussed in Chapter III for the UPA for a central potential, this amounts merely to reading all the equations "the other way around". We can fit (Bhatt et al., 1972) the deuteron wave function for the Reid triplet soft core (1968) for instance by using (4.5) for the form factor $g_l(p)$; the strength λ in (4.7) is then adjusted so that $D(-B)=0$, where B is the deuteron binding energy.

Analytical expressions for the UPA form factors $g_0(p)$ and $g_2(p)$ were obtained (Bhatt et al., 1972) by choosing each as a sum of nine terms of Yamaguchi form [cf. (3.50)] and making a least squares fit to Reid's tabulation of his wave functions $u(r)$ and $w(r)$.

$$g_0(p) = \sum_{n=1}^{9} A_n/(p^2 + v_n^2),$$

$$g_2(p) = \sum_{n=1}^{9} B_n p^2/(p^2 + v_n^2). \tag{4.11}$$

The values used for the ranges v_n and for the coefficients A_n and B_n are given in Table 4.1.

Siebert et al. (1972) calculated the UPA t-matrix, $t_{LL'}^U(p, k; s)$ from (4.9); the denominator $D(s)$ is given in analytical form in Siebert's thesis (1973). The form factors $g_0(p)$ and $g_2(p)$ are plotted in Fig. 4.1. Note

Table 4.1. Parameters for UPA form factors for Reid triplet potential

n	Central triplet $g_0(p)$		Tensor $g_2(p)$
	v_n (fm^{-1})	A_n	B_n
1	0.7	2.9456 − 2	− 9.8470 − 3
2	1.4	− 2.5621 0	3.7037 − 1
3	2.1	7.1191 1	− 2.0821 1
4	2.8	− 5.7144 2	1.3012 2
5	3.5	2.0488 3	− 3.7884 2
6	4.2	− 2.6609 3	3.6828 2
7	5.6	1.5413 3	− 8.3648 1
8	8.4	− 5.3443 2	− 3.4179 1
9	12.6	1.0926 2	1.9866 1

For use in Eq. (4.11); from Bhatt et al. (1972).

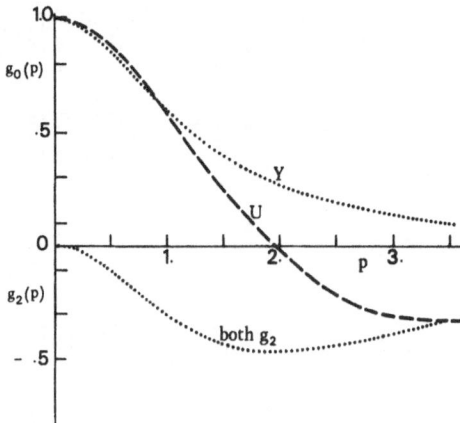

Fig. 4.1. Form factors for triplet potentials vs. momentum p in fm^{-1}. The dotted curve Y is the Yamaguchi central form factor; the dashed curve U is the central UPA for the Reid triplet; the dotted curve "both" represents both Yamaguchi and UPA for the tensor form factor — they disagree at much higher values of momentum p (in fm^{-1}). From Siebert et al. (1972)

that $g_0(p)$ is very different from the Yamaguchi central form factor, while $g_2(p)$ is very close to the Yamaguchi tensor form factor.

Siebert (1972) compares the UPA t-matrices, $t^U_{LL'}(p, k; s)$ with Haftel's numerical results (private communication) for the Reid triplet soft core potential, $t_{LL'}(p, k; s)$, and with the Yamaguchi $t^Y_{LL'}(p, k; s)$, for 7% D-state. See Table 4.2. Figure 4.2 compares "elements" for the three t-matrices, for $L = L' = 0$, and for $p = 1.0219\ \text{fm}^{-1}$ for variable k and three different negative values of energy s. We see that the UPA is quite accurate at the small negative energy of $s = -0.25\ \text{fm}^{-2}$, while the Yamaguchi form factor gives poor results for $k > 1.5\ \text{fm}^{-1}$. The accuracy of the UPA decreases as s becomes more negative but it is always better than a Yamaguchi shape. Figure 4.3 compares the values of t_{02}, t_{20}, and t_{22} for $p = 1.0219\ \text{fm}^{-1}$ and $s = 1.0\ \text{fm}^{-2}$. Finally Fig. 4.4 compares diagonal ($p = k$) elements of the three t-matrices again for $s = 1\ \text{fm}^{-2}$. Note that $t^U_{00}(p, p; s)$ cannot change sign; but it touches zero near the value of p at which Haftel's values change sign.

Afnan and Read (1973A) have also studied the accuracy of the UPA fit to the t-matrix for the Reid triplet potential. Their Table 2 agrees with our Table 4.2 to an accuracy of several percent. (Afnan denotes our UPA as "UPAI".)

We now consider on-shell matrix elements: i.e., the phase parameters for the coupled system. Fuda (1968A) gave a simple solution using

Table 4.2. Comparison of t-matrices for energy of $-1\,\mathrm{fm}^{-2}$ for Reid triplet

p	k	t_{oo}	t_{oo}^U	t_{oo}^Y	t_{02}	t_{02}^U	t_{02}^Y	t_{20}	t_{20}^U	t_{20}^Y	t_{22}	t_{22}^U	t_{22}^Y
0.2	0.2	-1.962	-2.000	-1.936	0.084	0.039	0.037	0.084	0.039	0.037	0.003	-0.001	-0.001
0.2309	0.2	-1.945	-2.008	-1.919	0.081	0.038	0.037	0.111	0.052	0.049	0.004	-0.001	-0.001
0.5743	0.2	-1.642	-1.730	-1.635	0.035	0.033	0.032	0.487	0.266	0.266	0.008	-0.005	-0.005
1.0219	0.2	-1.093	-1.166	-1.182	0.011	0.023	0.023	0.858	0.604	0.614	0.003	-0.012	-0.012
1.7095	0.2	-0.221	-0.268	-0.684	0.002	0.005	0.013	1.048	0.901	0.893	0.003	-0.017	-0.017
2.9893	0.2	0.630	0.617	-0.291	-0.003	-0.012	0.006	0.851	0.831	0.762	0.004	-0.016	-0.014
0.2309	0.5	-1.709	-1.792	-1.694	0.391	0.210	0.208	0.057	0.046	0.044	0.010	-0.005	-0.005
0.5	0.5	-1.541	-1.611	-1.507	0.248	0.189	0.185	0.248	0.189	0.185	0.030	-0.002	-0.002
0.5743	0.5	-1.476	-1.543	-1.443	0.210	0.181	0.177	0.315	0.238	0.235	0.033	-0.027	-0.029
1.0219	0.5	-0.991	-1.041	-1.043	0.073	0.122	0.128	0.697	0.539	0.541	0.018	-0.063	-0.067
1.7095	0.5	-0.192	-0.239	-0.604	0.011	0.028	0.074	0.955	0.804	0.788	-0.019	-0.094	-0.097
2.9893	0.5	0.600	0.550	-0.257	-0.020	-0.065	0.032	0.805	0.742	0.672	-0.024	-0.087	-0.083
0.2309	1.0	-1.115	-1.188	-1.193	0.834	0.584	0.593	0.016	0.031	0.031	0.005	-0.015	-0.015
0.5743	1.0	-0.978	-1.023	-1.016	0.627	0.503	0.505	0.101	0.158	0.165	0.024	-0.078	-0.082
1.0	1.0	-0.685	-0.708	-0.747	0.309	0.348	0.371	0.309	0.348	0.371	0.024	-0.171	-0.185
1.0219	1.0	-0.668	-0.690	-0.735	0.295	0.340	0.365	0.321	0.357	0.381	0.022	-0.176	-0.189
1.7095	1.0	-0.099	-0.158	-0.425	0.046	0.078	0.211	0.659	0.533	0.555	-0.057	-0.262	-0.276
2.9893	1.0	0.508	0.365	-0.181	-0.069	-0.179	0.090	0.656	0.492	0.473	-0.080	-0.242	-0.235
0.2309	1.5	-0.467	-0.517	-0.798	1.017	0.838	0.841	0.005	0.013	0.021	-0.003	-0.022	-0.022
0.5743	1.5	-0.404	-0.445	-0.680	0.876	0.722	0.716	0.030	0.069	0.111	-0.014	-0.111	-0.117
1.0219	1.5	-0.258	-0.300	-0.492	0.566	0.487	0.517	0.097	0.155	0.255	-0.034	-0.252	-0.269
1.5	1.5	-0.060	-0.134	-0.335	0.226	0.217	0.353	0.226	0.217	0.353	-0.069	-0.352	-0.370
1.7095	1.5	0.029	-0.069	-0.285	0.119	0.112	0.300	0.295	0.232	0.371	-0.092	-0.375	-0.391
2.9893	1.5	0.388	0.158	-0.121	-0.124	-0.257	0.128	0.447	0.213	0.318	-0.134	-0.348	-0.334

0.2309	2.0	0.081	0.035	−0.546	1.042	0.934	0.900	−0.000	−0.001	0.014	−0.006	−0.024	−0.023
0.5743	2.0	0.089	0.033	−0.465	0.941	0.805	0.768	−0.001	−0.005	0.075	−0.034	−0.124	−0.124
1.0219	2.0	0.109	0.021	−0.336	0.705	0.543	0.554	0.001	−0.011	0.175	−0.083	−0.281	−0.288
1.7095	2.0	0.173	0.005	−0.194	0.239	0.125	0.320	0.035	−0.016	0.254	−0.134	−0.419	−0.419
2.0	2.0	0.192	−0.001	−0.157	0.075	−0.017	0.258	0.075	−0.017	0.258	−0.149	−0.438	−0.426
2.9893	2.0	0.274	−0.010	−0.082	−0.165	−0.286	0.136	0.216	−0.015	0.216	−0.161	−0.387	−0.357
0.2309	3.0	0.630	0.615	−0.287	0.845	0.823	0.752	−0.005	−0.016	0.007	−0.005	−0.021	−0.019
0.5743	3.0	0.592	0.530	−0.244	0.786	0.709	0.641	−0.026	−0.082	0.040	−0.031	−0.109	−0.104
1.0219	3.0	0.504	0.357	−0.177	0.646	0.478	0.464	−0.072	−0.184	0.092	−0.082	−0.247	−0.241
1.7095	3.0	0.339	0.082	−0.102	0.350	0.110	0.268	−0.145	−0.276	0.133	−0.149	−0.369	−0.350
2.9893	3.0	0.139	−0.189	−0.043	−0.142	−0.253	0.114	−0.152	−0.255	0.113	−0.125	−0.340	−0.299
3.0	3.0	0.131	−0.189	−0.043	−0.151	−0.254	0.114	−0.151	−0.254	0.114	−0.124	−0.339	−0.300

Exact $t_{LL'}(p,k;-1)$ calculated by Haftel for the Reid triplet potential. $t_{LL'}^{U}$ gives the UPA values, 4.9; while $t_{LL'}^{Y}$ gives values for Yamaguchi shapes with 7% D-state. From Siebert et al. (1972).

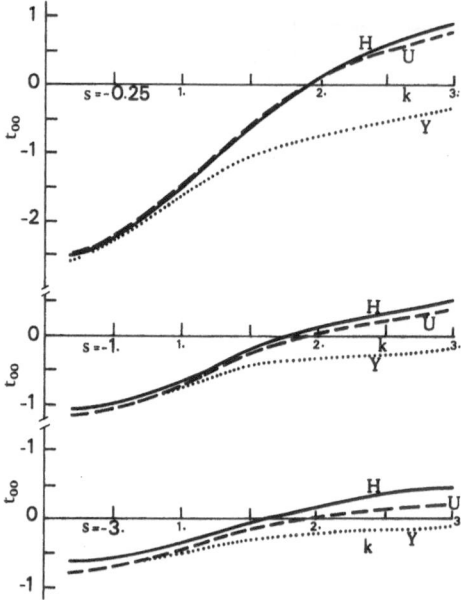

Fig. 4.2. Comparison of $t_{00}(1.0219, k; s)$ for energy $s = -0.25$, -1.0, and $-3.0\,\mathrm{fm}^{-2}$. The solid curves show Haftel's exact values for the Reid triplet; the dashed curves show UPA values, and the dotted curves show values for a Yamaguchi triplet potential. From Siebert et al. (1972)

Blatt-Biedenharn (1952) phase parameters,

$$(-ik + k\cot\delta_\alpha)^{-1} = 2\pi^2 [g_0^2(k) + g_2^2(k)]\, D_T^{-1}(k^2 + i\varepsilon)\,, \qquad (4.12)$$

$$\tan\varepsilon(k) = g_2(k)/g_0(k)\,, \qquad (4.13)$$

$$\delta_\beta(k) = 0\,, \quad \text{all } k\,. \qquad (4.14)$$

These equations can be derived in a straightforward manner by writing down the S-matrix and corresponding t-matrix in Blatt-Biedenharn notation, and comparing with (4.9) for the t-matrix in the unitary pole approximation. Equation 4.14 follows from the simple consequence of (4.9) that the mixed $(L \neq L')$ element of the on-shell t-matrix is the geometric mean of the diagonal t_{00} and t_{22}.

$$[t_{02}(k, k; s)]^2 = t_{00}(k, k; s)\, t_{22}(k, k; s)\,. \qquad (4.15)$$

In comparing Eqs. (4.12) to (4.14) with experiment, we must remember that published results generally give the Stapp barred phase shifts – and generally omit the bars! But one can convert from one to

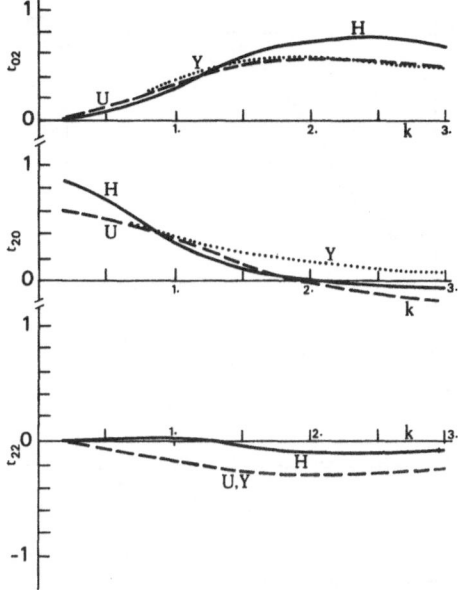

Fig. 4.3. Comparison of values for $t_{02}(1.0219, k; -1.0)$, $t_{20}(1.0219, k; -1.0)$, and $t_{22}(1.0219, k; -1.0)$. The solid curves show Haftel's exact values for the Reid potential, the dashed curves show UPA values, and the dotted curves show Yamaguchi values. From Siebert et al. (1972)

the other (Stapp et al., 1957). In particular,

$$\delta_\beta = \tfrac{1}{2}\{\bar{\delta}_0 + \bar{\delta}_2 - \sin^{-1}[\sin 2\bar{\varepsilon}(1 + x^2)^{\frac{1}{2}} x^{-1}]\} \tag{4.16}$$

where

$$x = \tan 2\bar{\varepsilon}/\sin(\bar{\delta}_0 - \bar{\delta}_2). \tag{4.17}$$

For instance at a laboratory energy of 200 MeV, MacGregor et al. (1969) give the Stapp barred phase shifts as $\bar{\delta}_0 = 16.89 \pm 0.58°$ (for the $^3S_1$ state); $\bar{\delta}_2 = -18.12 \pm 0.48°$ (for the $^3D_1$ state), and $\bar{\varepsilon} = 6.86 \pm 0.56°$ (for the mixing parameter). Equation (4.16) gives a Blatt-Biedenharn phase shift $\delta_\beta = 19.2 \pm 0.6°$ clearly different from the zero predicted by the UPA.

Previous papers (Siebert et al., 1972; Siebert et al., 1973; Afnan and Read, 1973A) reach the same conclusion that the UPA disagrees with experiment by comparing phase parameters in the Stapp barred representation. (This in turn leads to a disagreement among theorists as to the sign of the calculated Stapp mixing parameter.) The comparison above of calculated and experimental values of δ_β is particularly simple.

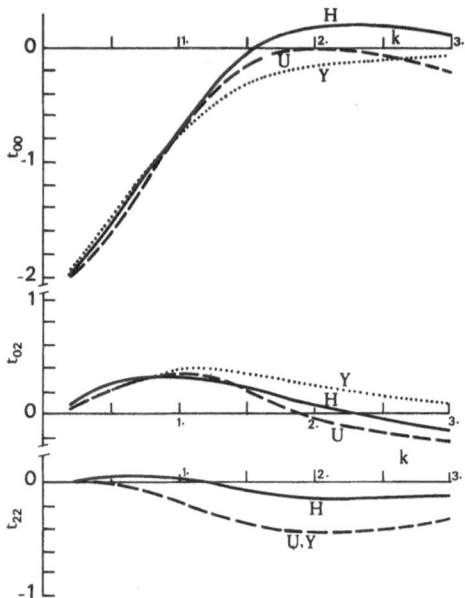

Fig. 4.4. "Diagonal" t-matrices $t_{LL'}(k, k; -1.0)$ vs. k. The solid curves show Haftel's values, the dashed curves show the UPA, and the dotted curves show Yamaguchi values. From Siebert et al. (1972)

In any case, the UPA clearly fails for the on-shell elements at positive energies.

This disagreement with experiment has recently been shown in still another, related, manner by Lambacher and Urban (1972), Pieper and Kowalski (1972) and Bhatt (1973). Namely, the UPA gives (4.15), which leads to a prediction of zero polarization P in nucleon-nucleon scattering in the coupled $^3S_1 - {}^3D_1$ states. But this prediction disagrees with experiment demonstrating a failure of the UPA. This recent work is equivalent to Fuda's earlier result (4.14), since Eqs. (4.15) and (4.14) can each be deduced from the other.

We therefore turn briefly to a simple modification of the UPA proposed by Afnan et al. (1971) and pursued by Afnan and Read (1973A). They replace the constant λ in (4.7) by three different constants, $\lambda_{LL'}$

$$v_{LL'}(p, k) = -\lambda_{LL'} g_L(p) g_{L'}(k). \tag{4.18}$$

Since the potential is Hermitean, $\lambda_{02} = \lambda_{20}$.

Afnan and Read find that the parameters should be given by

$$\lambda_{LL'} = \langle \chi_L | G_0(-B) V_{LL'} G_0(-B) | \chi_{L'} \rangle \left[\langle \chi_L | G_0(-B) | \chi_L \rangle \right]^{-1}$$
$$\left[\langle \chi_{L'} | G_0(-B) | \chi_{L'} \rangle \right]^{-1}. \tag{4.19}$$

Here

$$|\chi_L\rangle = (H_0 + B)|\psi_L\rangle, \qquad (4.20)$$

where $|\psi_L\rangle$ is the wave function for orbital angular momentum L. The $V_{LL'}$ in (4.19) is the Reid potential, or any other potential being approximated by a rank-one separable approximation of form (4.18).

The use of Eqs. (4.18) to (4.20) provides what Afnan and Read call the "1 A" or single attractive term approximation. It provides a much better fit to the Reid phase parameters (in Stapp notation) than is provided by the UPA (Siebert et al., 1972; Afnan-Read, 1973A). The restriction that the Blatt-Biedenharn $\delta_\beta = 0$ is now removed: e.g., at 200 MeV Afnan and Read find $\bar\delta_0 = 12°$, $\bar\delta_2 = -15°$, $\varrho = 0.47$ giving $\delta_\beta = 18°$. Further, use of the Afnan-Read form provides a much better fit to $t_{22}(k, k; -1.0)$ than that shown by the UPA fit in Table 4.2 or Fig. 4.4. For instance for $k = 2$, the exact, UPA and Afnan-Read values are -0.149, -0.438, and -0.133 respectively. But the fit to t_{00} and t_{02} at negative energies is in general not improved.

For purposes of computing the two-body t-matrix, the fit (4.18) is no harder to use than the UPA assumption (4.7). However, the freedom introduced by Afnan and Read does lead to appreciable extra complications in solution of the three-nucleon problem: complications similar to those for use of a rank-two non-central potential.

Siebert (1973) has recently shown that the success of the UPA for Tabakin's (1964) rank-two tensor potential is qualitatively the same as for the Reid local soft core tensor and spin-orbit potential. Again, the UPA is particularly good at fitting $t_{00}(s)$ at negative s. The fit is less satisfactory for $t_{02}(s)$, $t_{20}(s)$ and particularly $t_{22}(s)$ at negative s. The UPA fit to the phase parameters is unsatisfactory: this result can be understood from our argument above that the UPA result of zero for the Blatt-Biedenharn δ_β must agree poorly with Tabakin's result, since he achieved a fair fit to the experimental values, significantly different from zero.

The spin-orbit term in the Reid potential does not have a major effect in the failures of the UPA fit to the t-matrix; though it is clearly desirable to take it into account in a better manner than just including its effect in the deuteron wave function (Afnan-Read, 1973A). Thus the V_{LS} changes $v_{22}(p, k)$ but does not affect $v_{00}(p, k)$ or $v_{02}(p, k)$.

4.3 Rank-N Separable, Central Case

Now that we have seen limitations in the use of a one-term separable t-matrix, we are motivated to study the more complicated case, (1.3), of a rank-N separable t-matrix. This study has two main uses: it provides a convenient expression for the t-matrix, which has proved of great utility (Harms et al., 1970); and it gives a basis for understanding the success of the UPA. We consider a central potential with a single bound state, and emphasize fits to the t-matrix at negative energy.

The first use of rank-2 separable potentials was a natural extension of the pattern established by Yamaguchi for a one-term separable potential: choose a convenient mathematical expression for the form factor, and adjust parameters to fit nucleon-nucleon data. This procedure was initiated by Tabakin (1964, 1965) for tensor and central potentials respectively. [Brady (1970) modified the parameter values of Tabakin's tensor fit, to fit the deuteron binding energy.] Mongan (1968, 1969) fitted singlet phase shifts, triplet phase parameters, and the deuteron energy and quadrupole moment. I shall not attempt to give a good listing of references; but merely remark that a more systematic approach seems desirable; though I cannot 'exclude the possibility that a very good low rank separable potential has already been found by trial and error methods.

Following Harms (1970) we start with an expansion in Sturmian functions, also called a Weinberg series (Weinberg, 1963; Ball et al., 1968).

We perform this expansion using a relatively unfamiliar complete set of orthonormal functions: the set of bound states with different numbers of nodes, but all at the *same energy z*. These functions are solutions of the Schrödinger equation for potentials all of the same shape, but of *different strengths*, adjusted so that the energy eigenvalue remains fixed at the specified value. Consider for example the well known non-relativistic ns states in a Coulomb field, where the principal quantum number $n = 1, 2, 3, \ldots$. Ordinarily we use the same potential – e.g., for an electron in a hydrogen atom – and find an orthonormal set of wave functions $|\psi_n\rangle$; but for completeness we also need the positive energy eigenfunctions of ionized hydrogen, $|\psi_E\rangle$. [See our expansion for $\mathbf{G}(s)$ in the Low equation, giving (3.14).] In the Sturmian expansion we fix the energy z, say at -1 Rydberg. Our states $|\chi_m\rangle$ are the product of $(z - \mathbf{H}_0)$ and any of the following. i) the $1s$ eigenfunction of hydrogen; ii) the $2s$ wave function of singly ionized helium; iii) the $3s$ wave function of doubly ionized lithium, etc. That is, we are considering all S-states at the same energy of -1 Rydberg for the isoelectronic sequence H, He$^+$, Li^{++}, etc. (I have chosen a special case, where we find in nature examples of potentials with the same shape, with suitably adjusted strength.) We assert (Weinberg, 1963) that the discrete infinity of states $|\chi_m\rangle$ ($m = 1, 2, 3, \ldots$) form an orthonormal set in the sense of (4.23), and that the potential can be expanded in an infinite sum using the set $|\chi_m\rangle$. We then approximate by truncating the infinite series at N terms.

The potential is multiplied by the factor λ_m to keep the state with $(m-1)$ nodes in its radial function at energy z. We claim that the potential operator $\mathbf{V}$, can be expressed in terms of this set by the expansion

$$\mathbf{V} = \sum_{m=1}^{\infty} -\lambda_m^{-1} |\chi_m(z)\rangle \langle \chi_m(z)| \tag{4.21}$$

Harms' (1970) verification follows. The Schrödinger equation for $|\chi_n(z)\rangle$ giving a bound state at energy z for a potential $\lambda_n \mathbf{V}$ reads

$$|\chi_n(z)\rangle = \lambda_n(z)\, \mathbf{V}\, \mathbf{G}_0(z)\, |\chi_n(z)\rangle \,. \tag{4.22}$$

The orthonormality of the set $|\chi_n\rangle$ is expressed by

$$\langle \chi_m(z)| G_0(z) |\chi_n(z)\rangle = -\delta_{nm} \tag{4.23}$$

where the right side is the Kronecker δ symbol. (The minus sign is chosen because the Green's function for a free particle $\mathbf{G}_0(z)$ is negative for negative energy z.) Substituting (4.21) in (4.22) and using (4.23) we obtain an identity.

We truncate the expansion (4.21) at some finite N, thus expressing the $\mathbf{V}$ as a rank-N separable potential. The LS equation (3.11) is not much harder to solve for a rank-N separable potential than for a rank-1 separable which we used in Chapter III. We can sort out the N simultaneous algebraic equations, or we can introduce Stagat's notation (1969) in which the potential is an operator in an N-dimensional vector space. In either event, following the same argumentation that led to the solution (3.39), we obtain a rank-N separable t-matrix, of the form (1.3). In solving the LS equation for $t(s)$, many workers (Weinberg, 1963; Ball et al., 1968, etc.) have equated the two energies: i.e., $s = z$. Harms (1970) pointed out that are two advantages in keeping the energy z fixed, say at the energy of the bound state: $z = -B$. First, the form factors for the rank-N separable potential are energy-independent in Harms' scheme: this makes them more convenient, particularly if they are known only in numerical form. Second, setting $z = -B$, the first term in the series for the t-matrix is exactly the UPA, which we have learned already represents a good approximation to the exact t-matrix. Harms's series will presumably be a better approximation; this relation to the UPA suggests naming it the unitary pole expansion, or UPE. We have

$$t_{\mathrm{UPE}}(s) = \sum_{n,m=1}^{N} |\chi_n(-B)\rangle\, \Delta_{n,m}(s)\, \langle \chi_m(-B)| \,, \tag{4.24}$$

$$-[\Delta(s)^{-1}]_{n,m} = \lambda_n\, \delta_{n,m} + \langle \chi_n(-B)| G_0(s) |\chi_m(-B)\rangle \,. \tag{4.25}$$

Harms (1970) applies (4.24) to the Reid soft core singlet potential, setting the energy z at zero. Since this potential has both attractive and repulsive terms, we can obtain bound states (with different numbers of nodes) by multiplying the potential by either positive or negative values of λ. Harms finds the values of λ given in Table 4.3. He also uses the functions $|\chi_m\rangle$ in the momentum representation, and obtains numerical values for the matrix elements of the series (4.24) for different values of N. In Table 4.4 the column "1 A" is just the UPA of Table 3.3; the "1 A + 1 R" uses $N = 2$, utilizing $\lambda_1^A = 1.0819$ and $\lambda_1^R = -0.06286$. The next column includes eigenfunctions for two positive and one negative

Table 4.3. Eigenvalues for Reid singlet

Positive	Value	Negative	Value
λ_1^A	1.0819	λ_1^R	−0.06286
λ_2^A	8.27	λ_2^R	−0.252
λ_3^A	21.0	λ_3^R	−0.55
λ_4^A	38.0		

See Eq. (4.22) for eigenvalues. From Harms (1970).

Table 4.4. Separable and exact T matrices for the Reid singlet potential

p	k	1 A	1 A + 1 R	2 A + 1 R	2 A + 2 R	3 A + 2 R	Exact
0.004	0.004	−1.5391	−1.5388	−1.5580	−1.5579	−1.5700	−1.5908
0.153	0.004	−1.5147	−1.5143	−1.5313	−1.5312	−1.5414	−1.5561
0.153	0.153	−1.4908	−1.4902	−1.5052	−1.5051	−1.5137	−1.5244
0.567	0.004	−1.2566	−1.2551	−1.2518	−1.2515	−1.2464	−1.2311
0.567	0.153	−1.2367	−1.2348	−1.2319	−1.2314	−1.2272	−1.2161
0.567	0.567	−1.0259	−1.0199	−1.0205	−1.0187	−1.0208	−1.0324
1.405	0.004	−0.4481	−0.4435	−0.4098	−0.4088	−0.3971	−0.3985
1.405	0.153	−0.4410	−0.4352	−0.4054	−0.4040	−0.3941	−0.3945
1.405	0.567	−0.3658	−0.3471	−0.3529	−0.3470	−0.3519	−0.3457
1.405	1.405	−0.1305	−0.0716	−0.1306	−0.1108	−0.1222	−0.1112
3.084	0.004	0.5242	0.5324	0.5285	0.5302	0.5165	0.5179
3.084	0.153	0.5159	0.5262	0.5228	0.5252	0.5137	0.5152
3.084	0.567	0.4279	0.4612	0.4619	0.4720	0.4776	0.4825
3.084	1.405	0.1526	0.2571	0.2639	0.2980	0.3112	0.3400
3.084	3.084	−0.1785	0.0068	0.0060	0.0645	0.0491	0.0921
6.878	0.004	0.0979	0.1030	0.1027	0.1031	0.1066	0.1038
6.878	0.153	0.0964	0.1028	0.1026	0.1030	0.1060	0.1035
6.878	0.567	0.0800	0.1006	0.1006	0.1027	0.1012	0.0998
6.878	1.405	0.0285	0.0933	0.0938	0.1006	0.0971	0.0854
6.878	3.084	−0.0334	0.0815	0.0815	0.0931	0.0972	0.0699
6.878	6.878	−0.0062	0.0650	0.0650	0.0673	0.0663	0.0794

$t(p, k; -0.5)$ for Harms' unitary pole expansion (Harms, 1970). The column "1 A" gives the UPA; "1 A + 1 R" includes one negative eigenvalue from Table 4.3, etc.

value of λ, etc. The last column gives the "exact" numerical solution of the LS equation, also given in Table 3.3. The success of the UPE in fitting the exact t-matrix for the Reid singlet potential is illustrated in Figs. 3.9, 3.10, 3.11, and 3.12, where the dotted curves show the UPE for three terms: using the first two positive eigenvalues and the first negative eigenvalue given in Table 4.3. We see that in this case the convergence of the UPE is quite good: three terms fit the exact results for the t-matrix within 0.05 units, or less, even for values of momenta and energy where the UPA has errors larger than 0.3 units.

I should note that Table 4.4 shows we should use some discretion in applications of the UPE. Using alternatively attractive and repulsive terms, we have a series of alternating terms that tend to cancel. In some cases this leads to the result that one term (the UPA) is more accurate than two terms in fitting the exact value.

Kok (1970) has applied the UPE to other potentials. He has also studied the possibility of fixing the energy z in the Sturmian expansion at a value different from $z = -B$ chosen by Harms. In general the UPE converges rapidly for central potentials; and the results are not sensitive to Kok's choice of z, provided it is not very different from Harms' choice.

I turn briefly to the use of rank-N separable t-matrices to fit the exact t-matrix for a central potential at positive energies s. Of course the UPE can be used here as well (Brady et al., 1972). Use of a higher rank separable potential allows the observed sign change of the singlet phase shift. The convergence of the UPE at positive energies does not seem to be as rapid as that illustrated above for negative energies.

Several other series have been studied for expansions of a local central potential as a sum of separable terms: the Fiedeldey (1969) fit using the phase shift, as well as the bound state eigenfunction; the fit by Ernst et al. (1973) using the eigenfunctions at two or more energies; and Osborn's (1969) generalization of the Noyes-Kowalski approximation.

Fiedeldey (1969) uses a rank-2 separable potential with form factors chosen to provide a fit to *both* the phase shifts *and* the bound state wave function for a specified local potential. The procedure consists of first fitting an attractive rank-1 separable potential to the bound state wave function (i.e., the UPA). He then finds another separable potential with form factor orthogonal to the UPA form factor, chosen to produce a fit to the phase shifts. This choice can be made provided the difference between UPA and desired phase shifts always has the same sign. But this provision is not met, for instance for the Reid singlet. (Private communications from E. Harms and H. Fiedeldey.)

One would expect that Fiedeldey's procedure should give quite a good fit to the t-matrix for a large range of momenta and energies, since it gives an exact fit both on the plane $(s = -B)$ and for the on-shell curve shown in Fig. 3.1.

Ernst et al. (1973) generalize the UPA by giving a rank-N separable potential that fits the eigenfunctions (either bound state or continuum) at N different energy values. Thus in a two-term generalization of the UPA, one energy would be chosen at the bound state, and the other at a "suitable" positive energy state, E_1. Thus Ernst fits the exact t-matrix on two pk planes in Fig. 3.1 at $s = -B$ and at $s = E_1$.

Numerical comparisons of the UPE, Fiedeldey and Ernst fits for the same potential would be of interest.

Osborn's (1969) generalization of the Noyes (1965), Kowalski (1965) approximation uses a series of energy-dependent form factors to give a rank-N separable potential. In keeping with my arbitrary restriction to energy-independent potentials, I will not pursue Osborn's interesting work.

4.4. Rank-N Separable, Non-Central Case

The use of a rank-N separable approximation to the t-matrix for a non-central potential is a more complicated and less developed subject than that of a tensor UPA, or a rank-N separable potential, treated in Sections 4.2 and 4.3. I will only outline the work in this area.

First, Tabakin (1964), Mongan (1968, 1969) among others have given explicit forms of rank-2 separable tensor potentials. While they are not the result of a systematic method of fitting the t-matrix for some assumed potential, Mongan's potentials do a good job of fitting experimental data for on-shell values of the t-matrix. Thus they represent a specific prescription to go from on-shell to needed off-shell values. Mongan does not attempt to fit the t-matrix for some specified phenomenological potential, such as Reid's triplet soft core: instead Mongan's approach is in the spirit of Baranger (1969) and Sauer (1973) of extrapolating directly from on-shell to off-shell elements of the t-matrix. In Chapter 2 I discussed the low percentage D-state of the deuteron for Mongan's potential, and the possibility of checking Mongan's extrapolation by experiments on polarization effects in electron-deuteron elastic scattering. I turn to attempts to fit the t-matrix for the Reid triplet soft core.

The unitary pole expansion of Section 4.3 could be applied to a non-central potential: the functions such as $|\chi_m(-B)\rangle$ in 4.24 should have both an S-wave and a D-wave for calculations on the coupled $^3S_1 - {}^3D_1$ system. I have not seen numerical work on these fits.

In Section 4.2 I discussed the Afnan-Read (1973A) modification of the UPA, to allow for the variation of the parameter $\lambda_{LL'}$ with orbital

Table 4.5. T-matrices for Reid triplet potential

p (fm^{-1})	$t_{00}(p,p; -1.0)$					$t_{02}(p,p; -1.0)$					$t_{22}(p,p; -1.0)$				
	Exact	9A+6R	2A+1R	1A	UPAI	Exact	9A+6R	2A+1R	1A	UPAI	Exact	9A+6R	2A+1R	1A	UPAI
0.2	-1.958	-1.952	-1.969	-1.951	-2.029	-0.085	-0.081	-0.052	-0.045	-0.039	0.003	0.002	-0.000	-0.000	-0.001
0.5	-1.537	-1.536	-1.553	-1.552	-1.614	-0.248	-0.254	-0.234	-0.221	-0.190	0.030	0.029	-0.003	-0.007	-0.022
1.0	-0.682	-0.682	-0.670	-0.686	-0.714	-0.308	-0.313	-0.366	-0.402	-0.345	0.024	0.021	-0.038	-0.051	-0.167
1.5	-0.059	-0.058	-0.086	-0.135	-0.140	-0.225	-0.228	-0.238	-0.257	-0.221	-0.069	-0.071	-0.107	-0.106	-0.348
2.0	0.194	-0.194	-0.120	-0.000	-0.000	-0.075	-0.079	-0.061	-0.016	0.014	-0.148	-0.150	-0.150	-0.133	-0.439
3.0	0.132	-0.133	0.047	-0.175	-0.182	0.151	0.149	0.198	0.284	0.244	-0.124	-0.125	-0.090	-0.099	-0.327

Comparison of diagonal $t_{LL'}(p, k; s)$ for momenta $p = k$, and energy $s = -1.0\,\mathrm{fm}^{-2}$. The UPAI is the UPA of Table 4.2, the "1A" is a single term for potential 4.18; "2A + 1R" means use of two positive and one negative eigenvalue, etc. From Afnan and Read (1973A).

angular momenta L and L'. Their "1A" fit to the Reid triplet represents an improvement over the UPA, particularly for fits to $t_{22}(p, k; s)$ at negative s, and for the phase parameters ($\bar{\delta}_2$ or δ_β in Stapp or Blatt-Biedenharn notation). Afnan and Read use an expansion in which the lowest term is their "1A" term. I copy in Table 4.5 their Table 2 for diagonal matrix elements the Reid triplet, for energy $s = -1\,\mathrm{fm}^{-2}$. [In comparison with Table 4.2 from Siebert et al. (1972), the reader should note that there is a different sign convention for $t_{02} = t_{20}$. There are also small numerical differences: e.g., at the upper left in Table 4.2, $t_{00}(0.2, 0.2, -1.0) = -1.962$; this matrix element is given as -1.958 in Table 4.5, also at the upper left.] Table 4.5 shows that the "2A + 1R" approximation gives rather good fits to the exact values of the Reid t-matrix. This approximation corresponds to the 3-term UPE given in Table 4.4 (same notation) and Figs. 3.9 to 3.12 for the Reid singlet potential, and has a similar accuracy. (One should compare the difference between the "2A + 1R" column and the exact values, *not* the percentage difference, since it is the difference which affects the triton energy.)

Fuda (1970) generalized Fiedeldey's two-term fit for the central case to give a rank-2 separable tensor potential to fit a given eigenfunction and set of phase parameters. The method may fail for the Reid triplet soft core, just as Fiedeldey's two-term fit fails for the Reid singlet soft core.

The technique of Ernst et al. (1973) applies in principle to a noncentral case, as well as the central case discussed briefly above.

Fuda (1969) has also generalized Osborn's (1969) series to give an (energy-dependent) rank-N separable non-central fit to the t-matrix for a specified non-central potential.

The Afnan-Read series has been pursued by them in detail, and also in application to the triton energy. At present it seems the most promising rank-N separable approximation for a t-matrix for the Reid triplet or similar local potential.

V. Solution of Trinucleon for Separable t-Matrix

5.1 Introduction

In this chapter I present the Faddeev equations for the trinucleon (^{3}H or ^{3}He). I give details of their derivation, and of the solution of the resulting one-dimensional integral equation for a central, spin-independent rank-one separable potential acting only in two-body S-states. This oversimplified model provides a check on the separable approximation applied to the calculation of the trinucleon energy, E_T, since we can compare with other calculations (Chapter VI) for corresponding local two-body potentials. I then quote without proof the form the Faddeev equations take for a separable tensor potential. I discuss the results for E_T using Yamaguchi shapes for the three form factors, and also using the unitary pole approximation discussed in Chapters III and IV for the Reid singlet and triplet soft core potentials. I also quote the equations we need to solve for rank-N separable potentials, and the resulting values of E_T when we use the unitary pole expansion given in Chapter IV for the Reid singlet potential or the Afnan-Read form for the Reid triplet potential.

I want to emphasize that the Faddeev equations (Faddeev, 1960; Watson et al., 1967) are nothing more than a convenient way of re-writing the Schrödinger equation for a three-body system. Indeed, if we assume a rank-one separable potential, it is possible to solve the Schrödinger equation directly. Mitra (1962, 1969) originally used this latter procedure. But using the Faddeev formulation has four significant virtues. First, we avoid the disconnected diagrams that plague three-body and many-body problems (Watson, 1967, Section 4.2). Second, we establish that the solution of the three-body problem depends just on the off-shell two-body t-matrices, rather than on the potential. (Hence our insistence above that we were interested in approximations to the t-matrix, *not* to the potential.) We also learn the range of values of momenta and of energy for which we need an accurate approximation to **t**. Third, we can use the Faddeev formulation for a variety of problems which would be inconvenient to treat without it: e.g., the trinucleon for a local potential, treated in the next chapter, or nucleon-deuteron scattering problems discussed in the last chapter. Fourth, we have the basis to develop t-matrix perturbation theory, used in the following chapter.

5.2 Faddeev Equation

We begin with the usual definition of coordinates for a three-body problem. For simplicity we treat the case where each nucleon has unit mass; but the formalism is applicable to the case where the three masses are different (Watson, 1967, Section 4.1).

We work in the center of mass system of the three nucleons. The nucleon coordinates are r_1, r_2, and r_3; and the momenta are p_1, p_2, and p_3. Of course, we need only two vectors to specify the nucleon coordinates, and two others for the canonically conjugate momenta. We choose the coordinate vectors r and ϱ illustrated in Fig. 5.1, defined as follows.

$$r = r_1 - r_2 ; \quad \varrho = 3 r_3/2 . \tag{5.1}$$

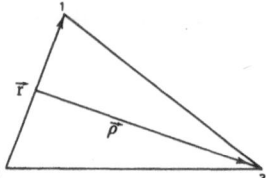

Fig. 5.1. Coordinate system used for trinucleon calculation: see Eq. (5.1)

The canonically conjugate momenta are

$$p = \tfrac{1}{2}(p_1 - p_2) ; \qquad q = p_3 . \tag{5.1'}$$

In words, r is the vector from the second to the first nucleon; and ϱ is the vector from the center of mass of the first two nucleons to the third nucleon. Using these coordinates and $\hbar = M = 1$, the three-body kinetic energy operator $\mathbf{H}_0^{(3)}$ is

$$\mathbf{H}_0^{(3)} = p^2 + 3q^2/4 . \tag{5.2}$$

We assume that the potential energy is the sum of two-body potentials, V_α; where the subscript indicates the nucleon that is *not* in the interacting pair: e.g., $V_3(r)$ is a central force between the first and second nucleons. (Three-body forces could be included: Lovelace, 1964.) The Schrödinger equation for trinucleon ket $|\psi\rangle$ for a bound state at negative energy E_T is

$$|\psi\rangle = (E_T - \mathbf{H}_0^{(3)})^{-1} \sum_{\alpha=1}^{3} V_\alpha |\psi\rangle . \tag{5.3}$$

We use the notation

$$|\psi_\alpha\rangle = (E_T - \mathbf{H}_0^{(3)})^{-1} V_\alpha |\psi\rangle . \tag{5.4}$$

Then the complete ket on the left of (5.3) can be written

$$|\psi\rangle = |\psi_\alpha\rangle + |\psi_\beta\rangle + |\psi_\gamma\rangle . \tag{5.5}$$

Substituting (5.5) in (5.4), and rearranging terms, we find

$$[1 - (E - \mathbf{H}_0^{(3)})^{-1} V_\alpha]|\psi_\alpha\rangle = (E - \mathbf{H}_0^{(3)})^{-1} V_\alpha [|\psi_\beta\rangle + |\psi_\gamma\rangle] . \tag{5.6}$$

Here the three subscripts α, β, and γ are a permutation of the integers, 1, 2, and 3.

We solve for the "pair ket" $|\psi_\alpha\rangle$ by multiplying by the inverse of the operator at the left of (5.6). We find this inverse by using a similar operator notation to that of Chapter III for the two-body t-matrix $t(s)$ in the Low equation (3.8) and the Lippmann-Schwinger equation (3.6). Our new operator $\mathbf{T}_\alpha(s)$ involves only the two-nucleon potential, but it includes the three-body kinetic energy operator $\mathbf{H}_0^{(3)}$. We write

$$
\begin{aligned}
\mathbf{T}_\alpha(s) &= V_\alpha + V_\alpha \mathbf{G}_\alpha^{(3)}(s)\, V_\alpha = V_\alpha + V_\alpha \mathbf{G}_0^{(3)}(s)\, \mathbf{T}_\alpha(s) \\
&= V_\alpha + \mathbf{T}_\alpha(s)\, \mathbf{G}_0^{(3)}(s)\, V_\alpha \,.
\end{aligned}
\tag{5.7}
$$

The three-body Green's function $\mathbf{G}_\alpha^{(3)}(s)$ for Hamiltonian $\mathbf{H}_\alpha^{(3)} = \mathbf{H}_0^{(3)} + V_\alpha$ is

$$
\mathbf{G}_\alpha^{(3)}(s) = 1/(s - \mathbf{H}_\alpha^{(3)}) \,.
\tag{5.8}
$$

The three-body Green's function for a free particle, with the Hamiltonian $\mathbf{H}_0^{(3)}$ of (5.2) is

$$
\mathbf{G}_0^{(3)}(s) = 1/(s - \mathbf{H}_0^{(3)}) \,.
\tag{5.9}
$$

I use the superscript 3 to emphasize that we are dealing with the kinetic energy for the three-nucleon system.

The following two operator equations can be verified easily using 5.7 to 5.9.

$$
[1 + \mathbf{G}_0^{(3)}(s)\, \mathbf{T}_\alpha(s)]\,[1 - \mathbf{G}_0^{(3)}(s)\, V_\alpha] = 1 \,,
\tag{5.10}
$$

$$
[1 + \mathbf{G}_0^{(3)}(s)\, \mathbf{T}_\alpha(s)]\, \mathbf{G}_0^{(3)}(s)\, V_\alpha = \mathbf{G}_0^{(3)}(s)\, \mathbf{T}_\alpha(s) \,.
\tag{5.11}
$$

We are now ready to solve (5.6) for $|\psi_\alpha\rangle$. We multiply on the left with the operator $[1 + \mathbf{G}_0^{(3)}(s)\, \mathbf{T}_\alpha(s)]$, and use (5.10) for the left side, and (5.11) for the right side. We find

$$
|\psi_\alpha\rangle = \mathbf{G}_0^{(3)}(s)\, \mathbf{T}_\alpha(s)\, [|\psi_\beta\rangle + |\psi_\gamma\rangle] \,.
\tag{5.12}
$$

These three equations (for the three choices of subscript) are the Faddeev equations for the trinucleon bound state, with energy $s = E_T < 0$.

Consider the case $\alpha = 3$. The operator $\mathbf{T}_3(s)$ involves only the potential between the first and second nucleon. It depends on energy s through the free three-particle Green's function, $\mathbf{G}_0^{(3)}(s)$ of (5.9). We can rewrite this dependence on s, including the momentum q^2 for the third nucleon: see Eq. (5.2) for $\mathbf{H}_0^{(3)}$. Thus we can express the "three-body" $\mathbf{T}_\alpha(s)$ in terms of the two-body t-matrix $t_3(s')$ at a different energy s'. We use the momentum representation, (5.1).

$$
\langle \mathbf{p}, \mathbf{q} \,|\, \mathbf{T}_3(s) \,|\, \mathbf{p}'\, \mathbf{q}' \rangle = \langle \mathbf{p} \,|\, t_3(s - 3q^2/4) \,|\, \mathbf{p}' \rangle\, \delta(\mathbf{q} - \mathbf{q}') \,.
\tag{5.13}
$$

The delta function shows that $\mathbf{T}_3(s)$ from (5.7), involves only the interaction of the first and second nucleons; the third (or spectator) nucleon does not change its momentum.

5.3 Separable Central Potential

Our results, (5.12) and (5.13) hold for any form of the two-body forces. I shall not develop the general form of the Faddeev equations in the momentum representation for a local potential (Ahmadzadeh et al., 1965; Harper et al., 1972), but merely quote the result when we need it in the next chapter. I shall instead express these equations in their simplest possible form: I assume a separable spin-independent potential acting only in two-body S states, in which case the Faddeev equations reduce to the single integral equation (5.22) in a single variable, as follows. We rewrite (5.12) for $\alpha = 3$ and energy $s = E_T$ using (5.13), (5.9) and (5.2). We look for a solution with a spatially symmetric S state. That is, the three kets $|\psi_1\rangle$, $|\psi_2\rangle$, and $|\psi_3\rangle$ in (5.12) are identical; but when we use a representation we must be careful to express each ket $|\psi\rangle$ in terms of appropriate variables.

$$\langle p, q | \psi_3 \rangle \equiv \psi_3(p, q)$$
$$= (E_T - p^2 - \tfrac{3}{4}q^2)^{-1} \int \langle p, q | T_3(E_T - \tfrac{3}{4}q^2) | p', q' \rangle \qquad (5.14)$$
$$\cdot [\langle p', q' | \psi_1 \rangle + \langle p' q' | \psi_2 \rangle] \, d^3 p' \, d^3 q' \,.$$

We use the Dirac delta function in (5.13) to perform the integral over $d^3 q'$. We write the variables in $\langle p' q | \psi_1 \rangle$ and in $\langle p' q | \psi_2 \rangle$ in terms of p' and q. Thus $|\psi_1\rangle$ is expressed in terms of $\tfrac{1}{2}(p_2 - p_3)$ and p_1. Using the definitions (5.1) we have $\tfrac{1}{2}(p_2 - p_3) = -\tfrac{1}{2}p - \tfrac{3}{4}q$ and $p_1 = p - \tfrac{1}{2}q$. We use these relations, and analogous ones for $|\psi_2\rangle$ to write

$$\langle p' q | \psi_1 \rangle = \psi(-\tfrac{1}{2}p' - \tfrac{3}{4}q \, ; \, p' - \tfrac{1}{2}q) \,,$$
$$\langle p' q | \psi_2 \rangle = \psi(-\tfrac{1}{2}p' + \tfrac{3}{4}q \, ; \, -p' - \tfrac{1}{2}q) \,. \qquad (5.15)$$

We rewrite (5.14) in the simpler form

$$\psi(p, q) = (E_T - p^2 - \tfrac{3}{4}q^2)^{-1} \int t(p, p' \, ; E_T - \tfrac{3}{4}q^2) \, d^3 p'$$
$$\cdot [\psi(-\tfrac{1}{2}p' - \tfrac{3}{4}q \, ; \, p' - \tfrac{1}{2}q) + \psi(-\tfrac{1}{2}p' + \tfrac{3}{4}q \, ; \, -p' - \tfrac{1}{2}q)] \,. \qquad (5.16)$$

The two terms in the square brackets must, from symmetry, give the same integral. (We can verify this explicitly by letting $p' = k + \tfrac{1}{2}q$ in the first term and $p' = -k - \tfrac{1}{2}q$ in the second term.) Then,

$$\psi(p, q) = 2(E_T - p^2 - \tfrac{3}{4}q^2)^{-1} \int t(p, k + \tfrac{1}{2}q \, ; E_T - \tfrac{3}{4}q^2)$$
$$\cdot \psi(\tfrac{1}{2}k + q, k) \, d^3 k \,. \qquad (5.17)$$

The integrand in (5.17) depends on two variables (for fixed vectors p and q): the magnitude k, and the angle θ between q and k. (This is an example of the two-dimensional integral equations found from the

Faddeev equations for a non-separable potential, acting only in relative S-states.)

We simplify further to a one-dimensional integral equation, using (3.39) for the two-body t-matrix for a rank-one separable potential with form factor $g(p)$.

$$
\psi(\boldsymbol{p}, \boldsymbol{q}) = - 2(E_{\mathrm{T}} - p^2 - \tfrac{3}{4}q^2)^{-1} g(p) [D(E_{\mathrm{T}} - \tfrac{3}{4}q^2)]^{-1}
$$
$$
\cdot \int g(|\boldsymbol{k} + \tfrac{1}{2}\boldsymbol{q}|)\, \psi(\tfrac{1}{2}\boldsymbol{k} + \boldsymbol{q}, \boldsymbol{k})\, \mathrm{d}^3 k \,.
$$
(5.18)

With our separable approximation, we have extracted all the dependence on momentum p from the integrand. The integrand depends only on the magnitude q, as shown below. We reduce (5.18) to a one-dimensional integral equation (5.22) by the following steps.

i) Define a *spectator function* that depends only on the magnitude of the momentum of the third, or spectator nucleon.

$$
\chi(q) \equiv 2[D(E_{\mathrm{T}} - \tfrac{3}{4}q^2)]^{-1} \int g(|\boldsymbol{k} + \tfrac{1}{2}\boldsymbol{q}|)\, \psi(\tfrac{1}{2}\boldsymbol{k} + \boldsymbol{q}, \boldsymbol{k})\, \mathrm{d}^3 k \,.
$$
(5.19)

ii) Express the wave function $\psi(\boldsymbol{p}, \boldsymbol{q})$ in terms of the spectator function, and substitute in (5.18), obtaining an integral equation for the spectator function:

$$
\chi(q) = 2[D(E_{\mathrm{T}} - \tfrac{3}{4}q^2)]^{-1} \int g(|\boldsymbol{k} + \tfrac{1}{2}\boldsymbol{q}|)\, g(|\tfrac{1}{2}\boldsymbol{k} + \boldsymbol{q}|)\, \chi(k)
$$
$$
\cdot (q^2 + \boldsymbol{q} \cdot \boldsymbol{k} + k^2 - E_{\mathrm{T}})^{-1}\, \mathrm{d}^3 k \,.
$$
(5.20)

iii) Write $\mathrm{d}^3 k = k^2\, \mathrm{d}k \sin\theta\, \mathrm{d}\theta\, \mathrm{d}\phi$. The integrand is independent of ϕ; and the integration over θ can be performed either analytically (for instance, for the Yamaguchi choice of the form factor) or numerically.

iv) Define the kernel $K_{E_{\mathrm{T}}}(q, k)$ as the result of the angular integration:

$$
K_{E_{\mathrm{T}}}(q, k) \equiv \iint g(|\boldsymbol{k} + \tfrac{1}{2}\boldsymbol{q}|)\, |g|(\boldsymbol{q} + \tfrac{1}{2}\boldsymbol{k}|)\, (q^2 + \boldsymbol{q} \cdot \boldsymbol{k} + k^2 - E_{\mathrm{T}})^{-1}
$$
$$
\cdot \sin\theta\, \mathrm{d}\theta\, \mathrm{d}\phi \,.
$$
(5.21)

v) We now have the desired one-dimensional integral equation for the spectator function:

$$
\chi(q) = 2[D(E_{\mathrm{T}} - \tfrac{3}{4}q^2)]^{-1} \int_0^\infty K_{E_{\mathrm{T}}}(q, k)\, \chi(k)\, k^2\, \mathrm{d}k \,.
$$
(5.22)

5.4 Numerical Methods and Results

The integral equation (5.22) is solved to find the negative trinucleon energy E_{T}. This was done independently over ten years ago by Kharchenko (1962) and by Mitra (1962). Here is a brief discussion of numerical methods. Equation (5.22) is simple enough that, with modern computers, we need not be careful about the details of our numerical techniques. But we shall later be interested in solutions of much harder equations: 1. three coupled integral equations of the form (5.22), for separable tensor forces; 2. a larger number of coupled integral equations, for rank-N separable tensor forces; 3. coupled two-dimensional integral equations for local potentials; 4. inhomogeneous equations for nucleon-deuteron scattering. So "details" can make the difference between an impossible and a possible calculation.

First, I note that using a computer, it costs almost as much time to use a Yamaguchi form for $g(p)$, as a more complicated analytical form: e.g., that of (3.55) found by Harms et al. (1970) for the Reid singlet.

Second, the desired eigenvalue E_T occurs non-linearly in (5.21) and (5.22). We avoid this complication by introducing a "fake" eigenvalue $\mu(E)$ which occur linearly:

$$\chi(q) = 2[D(E - \tfrac{3}{4}q^2)]^{-1} \mu(E) \int_0^\infty K_E(q, k)\, \chi(k)\, k^2\, \mathrm{d}k . \tag{5.23}$$

We solve (5.23) for different values of energy E, until we find a value E_T such that

$$\mu(E_T) = 1.0000\ldots \tag{5.24}$$

Third, we use a quadrature formula with N points to convert (5.23) into a matrix-vector equation, with N eigenvalues $\mu_n(E)$; $n = 1, 2, \ldots N$. We obtain better accuracy for a given value of N by using our knowledge of the asymptotic behavior of the integrand, both for k very small, and for k very large. Stagat (1968) and Brady et al. (1969) therefore use Gauss-Gegenbauer integration instead of the more usual gaussian integration.

Finally, one is tempted to use the usual Jacobi diagonalization of the matrix-vector problem to find *all* N eigenvalues $\mu_n(E)$. But we want only the largest eigenvalue, $\mu_N(E)$. We can obtain this single eigenvalue much more rapidly, by successive matrix multiplication, starting with an assumed eigenvector. After we obtain $\mu_N(E)$ for two initial values of E, we interpolate (or if necessary extrapolate) until we find E_T as the solution of (5.24).

The solution of (5.23) gives us the spectator function, which has two uses. First, the range in which the spectator function is appreciable gives us the range of momenta for which we need a good approximation to the t-matrix: cf. our estimate, Eq. (3.49). Second, we can use the spectator function (5.19) in (5.18) to find the trinucleon wave function $\psi(p, q)$ which can then be used to find other trinucleon properties: e.g., coulomb energy, or electromagnetic form factors for electron-trinucleon scattering (Gupta, 1965).

Of course, the use of a central, spin-independent separable potential is only a crude model to the nucleon-nucleon force we discussed in Chapter II. Still, this model does allow a check on the accuracy of the UPA by comparison of the value of E_T for the UPA with the value found by other means for the corresponding local potential (i.e., with the same two-body bound state ket $|B\rangle$). Kok et al. (1968) made this comparison for a local Hulthén potential, where the trinucleon energy was found from a variational calculation.

$$\text{Hulthén local (set 1): } -7.3 \text{ MeV},$$
$$\text{UPA for Hulthén (i.e., Yamaguchi): } -5.97 \text{ MeV}. \tag{5.25}$$

Harms et al. (1969) made this comparison for the Malfliet-Tjon (1969) spin-independent central potential (their number V) that consisted of one attractive and one repulsive local Yukawa potentials. Malfliet found the value of E_T by solution of a two-dimensional integral equation [see (5.17); the discussion in Chapter VI]. The results are:

$$\text{Malfliet local: } -7.3 \pm 0.1 \text{ MeV},$$
$$\text{UPA for Malfliet: } -7.44 \text{ MeV}, \tag{5.26}$$
$$\text{UPE for Malfliet: } -7.56 \text{ MeV}.$$

I include Harms' result (1970) using his unitary pole expansion for the MT potential, and extrapolating to an infinite number of terms in the UPE. (See below for discussion of the UPE.)

Brady et al. (1969) compare Tabakin's (1965) result for a central spin-independent rank-2 separable potential with the value of E_T for the UPA to this potential; and also with a Yamaguchi separable potential with the same effective range parameters (Tabakin, 1965).

$$\text{Tabakin two-term separable: } -8.40 \text{ MeV},$$
$$\text{UPA to Tabakin: } -8.60 \text{ MeV}, \tag{5.27}$$
$$\text{Yamaguchi: } -9.36 \text{ MeV}.$$

The comparison (5.25) shows that the UPA is only fair for the Hulthén local potential — the same conclusion reached in Table 3.1 and Fig. 3.2 to 3.5 for the t-matrix values at negative energy for a similar Yukawa potential. On the other hand, (5.26) and (5.27) show that the UPA works to an accuracy of (1/5) MeV for E_T for potentials with a "soft core", in accord with the good UPA fit to the t-matrix for the Reid soft core singlet potential: Table 3.3 and Figs. 3.8 to 3.12, Eq. (5.27) also gives Tabakin's (1965) result that the core raises the value of E_T by 1.0 MeV.

5.5 Non-Central Separable Potential

We omit discussion of spin-dependent *central* separable potentials, and go to the more realistic case of separable tensor potentials, studied by Bhakar and Mitra (1965) and by Sitenko et al. (1965); and later by Fuda (1968 A) and by Phillips (1968). All these potentials have the form $-\lambda g(p) g(k)$ proposed by Yamaguchi et al. (1954). The calculation is much more lengthy than that for central potentials, since we must consider different trinucleon states. We assume that isospin is exact. It turns out that we have three spectator functions $\chi_1(q)$, $\chi_2(q)$ and $\chi_3(q)$ which obey three coupled one-dimensional integral equations. In Fuda's notation (1968 A)

$$\chi_i(q) = F_i(q) \int d^3 k (q^2 + q \cdot k + k^2 - E_T)^{-1} \sum_{j=1}^{3} G_{ij}(q, k) \chi_j(k) . \tag{5.28}$$

$$\begin{aligned} F_1(q) &= F_2(q) = [2 D_{tr}(E_T - \tfrac{3}{4} q^2)]^{-1} , \\ F_3(q) &= [2 D_s(E_T - \tfrac{3}{4} q^2)]^{-1} . \end{aligned} \tag{5.29}$$

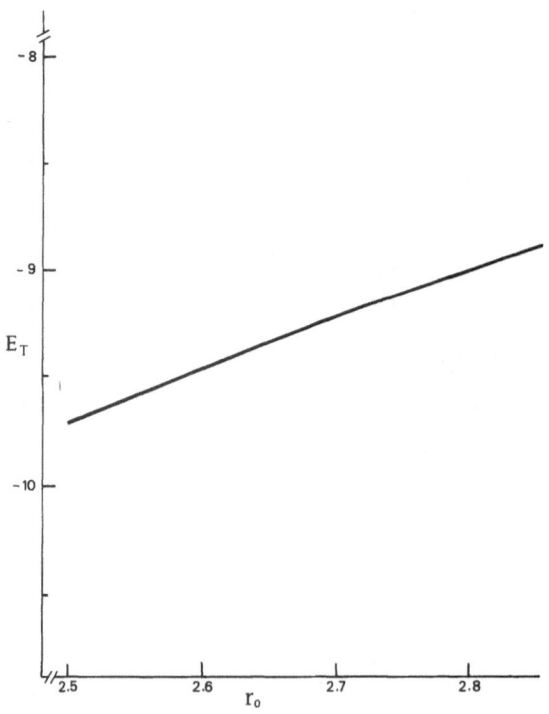

Fig. 5.2. Trinucleon energy E_T in MeV vs. singlet effective range r_0 in F, for separable potentials of Yamaguchi form. From Phillips (1968)

The kernels $G_{ij}(q, k)$ are given in Fuda's appendix. D_{tr} is the spin triplet denominator function, Eq. (4.10), while D_s is the spin-singlet denominator function, Eq. (3.38). These three coupled equations are solved by replacing them by a matrix-vector equation in $3N$ dimensional space. Using Gauss-Gegenbauer quadrature, a value of $N = 10$ is large enough, so the resulting eigenvalue problem is solved rapidly even on a modest computer to an accuracy of 0.01 MeV.

The calculated value of E_T depends on 1. the value used for the singlet effective range, 2. the value used for the deuteron percentage D-state, p_D, and 3. the shapes used for the separable form factors (singlet, central triplet, tensor triplet). It seems likely that a separable approximation is good enough for a semi-quantitative study of these three effects; though the UPA is not completely satisfactory for non-central forces or for a purely attractive central force.

Figures 5.2 and 5.3 illustrate the first two effects, as calculated by Phillips (1968) and by Brady et al. (1969). We see that a change of the singlet effective range of 0.15 F changes E_T by over $\frac{1}{3}$ MeV; and a

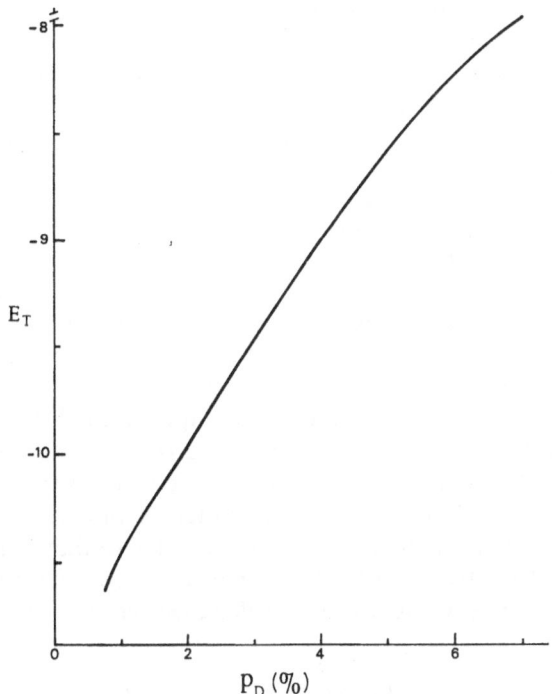

Fig. 5.3. Trinucleon energy E_T in MeV vs. deuteron percentage D-state, p_D, for separable potentials of Yamaguchi form. From Brady et al. (1969)

change of $1\frac{1}{2}\%$ in p_D gives a $\frac{1}{2}$ MeV change in E_T. In studying the third effect, we should clearly compare potentials with the same value of r_0 and p_D. A variety of such comparisons have been made (Kharchenko et al., 1968; Brady et al., 1969). I shall quote only the comparison between Yamaguchi shapes for both singlet and triplet form factors (interpolated to $p_D = 6.4\%$ which agrees with Reid's value) and the UPA (Bhatt et al., 1972) for Reid soft core singlet and triplet potentials.

$$\text{Yamaguchi shapes, } E_T = -8.12 \text{ MeV}\,,$$
$$\text{UPA for Reid } E_T = -7.45 \text{ MeV}\,. \tag{5.30}$$

That is, the UPA for a soft core increases the triton energy by some $\frac{2}{3}$ MeV. [Compare Tabakin's 1965 result, (5.27), for spin-independent-central: the UPA to the "core" increased E_T by $\frac{3}{4}$ MeV; while Tabakin's calculation gave E_T one MeV larger than that for a Yamaguchi potential.]

5.6 Higher Rank Non-Central Potentials

Finally, I consider calculations with rank-N tensor separable potentials: i.e., with rank n_1 for the singlet, rank n_2 for the central triplet, and rank n_3 for the tensor. The formalism was developed by Harms (1970 B) in a concise form, and leads to equations somewhat more complex than (5.28) for $n_1 = n_2 = n_3 = 1$. We obtain $v = (n_1 + n_2 + n_3)$ coupled integral equations. These can be solved in a reasonable time on a moderate computer: e.g., by Brady (1970) for $v = 6$.

I consider three applications of this formalism: i) by Siebert et al. (1973) to test the accuracy of the UPA for the rank-two tensor Tabakin (1964) potential; ii) by Harms et al. (1970) to test the accuracy of the UPE for the Reid singlet; and iii) by Afnan and Read (1973 B) to test their generalized tensor separable potential, with coupling constant $\lambda_{LL'}$ dependent on the two angular momenta.

As discussed in Chapter IV, Siebert has applied the UPA formalism to find a rank-one separable tensor potential of form (4.6) that has the same "deuteron" wave function, and the same residue at the singlet anti-bound state, as Tabakin's rank-two (1964) tensor potential. The agreement between the two triplet t-matrices is similar to that found for the UPA fit to the Reid triplet potential. Siebert compares the triton energy calculated with the UPA to that calculated earlier by Brady for Tabakin's potential.

$$\text{UPA for Tabakin, } E_T = -7.34 \text{ MeV}\,,$$
$$\text{Tabakin potential, } E_T = -7.02 \text{ MeV}\,. \tag{5.31}$$

Table 5.1. Triton energy calculated with UPE for Reid singlet

Singlet terms	$E_T(p_D = 4\%)$	$E_T(p_D = 7\%)$
1 A (or UPA)	-8.182	-7.579
1 A + 1 R	-8.058	-7.509
2 A + 1 R	-8.111	-7.544
2 A + 2 R	-8.069	-7.520
3 A + 2 R	-8.105	-7.548
∞ A + ∞ R	-8.14 ± 0.05	—

From Harms et al. (1970). Energy of triton in MeV, as we increase the number of terms (attractive A, or repulsive R) in the unitary pole expansion for the singlet fit to the Reid singlet soft core potential. Results are given for two different values of deuteron percentage D-state. The extrapolation to an infinite number of terms, with estimated maximum error, is discussed by Harms et al. (1970).

That is, the UPA is accurate to $\frac{1}{3}$ MeV — not quite as accurate as the $\frac{1}{5}$ MeV achieved in 5.27 for Tabakin's (1965) rank-two spin-independent central potential.

On the other hand, the UPE converges quite rapidly for the Reid singlet potential, and the first term (the UPA) was shown by Harms to give a very good approximation to the triton energy. Harms' values of E_T, vs. increasing number of terms n_1 in the UPE for the Reid singlet, are given in Table 5.1, and illustrated in Fig. 5.4. (The triplet central is separable with "modified Hulthén" shape: see Brady et al., 1969. The triplet tensor has a Yamaguchi shape, with parameters chosen so $p_D = 4\%$ or 7%. The latter choice for the triplet form factors gives similar form factors and similar values for E_T to those for the UPA to the Reid soft core triplet: see Bhatt et al., 1972.) We confirm the rapid convergence of the UPE, found in Chapter IV for values of the t-matrix at negative energy. The UPA (-8.18 MeV) and the three-term UPE (-8.11 MeV) each gives a value of E_T within 0.1 MeV of the value Harms finds by extrapolating the UPA to an infinite number of terms for $p_D = 4\%$, namely $E_T = -8.14 \pm 0.05$ MeV. (The error represents an estimated maximum error in extrapolation.) Note that the error in the use of the UPA for the Reid singlet is an order of magnitude smaller than the uncertainty in the trinucleon energy due to the assumed value of the deuteron percentage D-state. (Also see Fig. 5.3 for this effect, and see Fig. 5.2 for the relatively large effect of changes in the value of the singlet effective range.)

The Afnan-Read separable potential is given as (4.18); it provides a better fit to the two-body t-matrix than that given by the UPA. However, replacing (4.6) by (4.18) leads to appreciably more work in solution of the three-body problem: e.g., Bhatt's UPA for the Reid potential gives

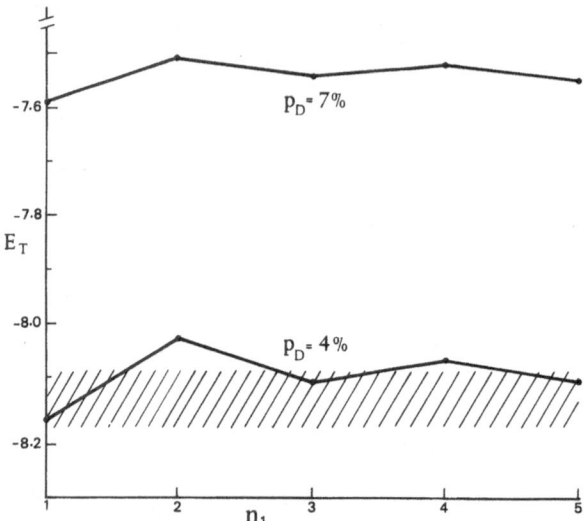

Fig. 5.4. Convergence for different number of terms n_1 in the unitary pole expansion for the Reid singlet potential. The trinucleon energy E_T in MeV is given for two different triplet potentials, with deuteron D-state of 4% and 7% respectively. The shaded region for the lower portion of the graph shows the extrapolated energy, and estimated maximum error, for an infinite number of terms in the UPE. See Table 5.1. From Harms et al. (1970)

three coupled one-dimensional integral equations, while the same UPA singlet and the Afnan-Read triplet gives five coupled one-dimensional integral equations. They also solve for higher rank separable tensor potentials given by their UPE development. I show Bhatt's UPA result, along with Afnan-Read using a single attractive term (1 A), one attractive and one repulsive term (1 A + 1 R), and two attractive and one repulsive term (2 A + 1 R) for the Reid triplet. In all cases the UPA is used for the Reid singlet. Other two body states (except the 1 S and coupled $^3S_1 - {}^3D_1$) are neglected in the present calculation.

$$
\begin{aligned}
\text{UPA for Reid triplet } E_T &= -7.45 \text{ MeV}, \\
\text{1 A for Reid triplet } E_T &= -7.15 \text{ MeV}, \\
(1\,\text{A} + 1\,\text{R}) \text{ for Reid triplet } E_T &= -6.99 \text{ MeV}, \\
(2\,\text{A} + 1\,\text{R}) \text{ for Reid triplet } E_T &= -7.04 \text{ MeV}.
\end{aligned}
\tag{5.32}
$$

This equation illustrates the change of $\frac{1}{3}$ MeV from the UPA value, due to use of the Afnan-Read single term. The latter gives a much better fit to the two-particle D-wave. The "1 A" term gives about as good a fit to the triplet as the UPA does for a central potential with a core: namely within $\frac{1}{5}$ or $\frac{1}{6}$ MeV in the triton energy.

5.7 Summary

The t-matrix for the Reid soft core potential can be well approximated by a low rank separable potential with a tensor force, for purposes of calculation of the triton energy. A single term is enough for the Reid singlet; the "single attractive term" of Afnan and Read is desirable for the Reid triplet fit. The accuracy of these low rank fits is shown by comparison with calculations using higher rank fits. In the next chapter the accuracy is re-examined, by using t-matrix perturbation theory, and by comparison with other calculations for the Reid potential.

VI. Survey of Trinucleon Calculations by Other Methods

6.1 Introduction

In the preceding Chapter I presented the methods and results for calculations of the trinucleon energy E_T using a separable approximation to the two-nucleon t-matrix. Here I treat briefly several other competing methods of calculating the same quantity: i) use of t-matrix perturbation theory; ii) solution of the two-dimensional integral equation (or coupled equations) resulting from the Faddeev equation for a non-separable t-matrix; iii) variational calculations; iv) use of expansions in hyperspherical harmonics and v) the Brueckner-Bethe method. I shall not attempt to give a detailed exposition of these methods. Instead I shall limit myself to indicating the mathematical methods, giving references for the details. I shall quote results emphasizing two choices of the two-body potential: 1. the central spin-independent potential V (CSI) of Malfliet-Tjon (1969) which is a combination of an attractive and a repulsive Yukawa potential; and 2. the non-central Reid soft core (1968) potential. In comparing results for different methods, I shall state whether higher partial waves in the two-nucleon interaction (beyond $^1S_1$ and coupled $^3S_1 - ^3D_1$) have been used: e.g., they were neglected in our work in Chapter V, leading to Eq. (5.17).

I also compare different calculated results for E_T for the Reid potential with the experimental value of -8.48 MeV, and discuss reasons for the disagreement of one to two MeV (the range being due to disagreements among the calculations). I discuss briefly Fabre's calculation (1972) for the Coulomb energy of ^{3}He, and the evidence for a small difference between neutron-neutron and nucleon proton-proton forces. Finally, I compare calculations of the electromagnetic form factors for electron-trinucleon elastic scattering with each other, and with experimental values.

6.2 *t*-Matrix Perturbation Theory

Fuda (1968 B) found the first order shift in the triton energy ΔE_T, due to a perturbation Δt in the two-nucleon *t*-matrix, for a central spin-independent potential (also see Alt et al., 1967). I shall not repeat the derivation but merely point out that starting from the Faddeev equation (5.12) one can reasonably expect that the first order change in the triton energy will be proportional to a small change Δt_α in the *t*-matrix, $T_\alpha(s)$. Fuda's general formula for the energy shift is

$$\Delta E = \sum_\alpha \left[\langle \phi | - \langle \phi_\alpha | \right] \Delta t_\alpha [| \phi \rangle - | \phi_\alpha \rangle] \qquad (6.1)$$

where $| \phi \rangle$ is the unperturbed triton ket, and $| \phi_\alpha \rangle$ is the "pair ket": cf. Eq. (5.4). In general 3-dimensional integrals are involved in evaluation of (6.1).

Fuda explains that this form of perturbation theory should work well even if there is a large change in the potential, provided there is a small change in the *t*-matrix. For instance, the "perturbation" in the potential might be a hard core in which case Rayleigh-Schrödinger perturbation theory clearly fails. But a hard core of small radius could give rather modest changes in the *t*-matrix in the range of values of energy and momenta that contribute appreciably to (6.1). A slightly less extreme example is given by the Malfliet-Tjon V potential (1969), or by the Reid singlet potential (1968). In these cases the specified local potential is *very* different from the separable potential used to find the UPA: see Table 3.3 for Reid, where the separable potential is negative, for $p < 2 \text{ fm}^{-1}$ and $k < 2 \text{ fm}^{-1}$, but the momentum space potential is positive in this range of momentum space. Nevertheless Δt and ΔE are very small, so (6.1) is useful despite the failure of Rayleigh-Schrödinger perturbation theory.

Recently Afnan et al., (1973 A) applied (6.1) to the Malfliet-Tjon potential, using Harms' (1970) UPE for the unperturbed and "exact" *t*-matrices. They expand the unperturbed *t*-matrix to N_1 terms and the "exact" to N terms, so the perturbation consists of $(N - N_1)$ separable terms. This technique is convenient, since it reduces the 3-dimensional integration of (6.1) to $(N - N_1)$ 1-dimensional integrals. They find that for $N = 11$ and $N_1 = 3$, convergence is achieved to an accuracy of a keV in the value of E_T. (The convergence is tested by varying both N and N_1.) Their result of $- 7.539 \text{ MeV}$ is given in Table 6.1, for this potential, including only the S-wave interaction. I include also (5.26) for the UPA, and Harms' (1970) value for the UPE. Note the excellent agreement between Afnan's value, and Harms' UPE value.

Recently Afnan and Read (1973 B) have used the same technique to find the triton energy for the Reid non-central potential. Here they

Table 6.1. Calculations of E_1 for Malfliet-Tjon potential

Method	S-waves only	All waves	Worker
Unitary Pole Approx.	−7.44	-—	Harms et al. (1969)
Unitary Pole Expans.	−7.56	-—	Harms et al. (1970)
t-matrix perturb.	−7.539	-—	Afnan (1973 A)
2-dim. Faddeev Eqs.	−7.3	−7.6	Malfliet et al. (1969)
			Malfliet (1969)
Variational	-—	−7.778	Erens et al. (1971)
Hyperspher. harm.	-—	−7.783	Erens et al. (1971)

E_T is the triton energy in MeV, for the fifth central spin-independent local potential of Malfliet et al. (1969). See Eq. (5.26) for the UPA and UPE results and discussion. See Figs. 6.4 and 6.5 for convergence of the calculation with hyperspherical harmonics.

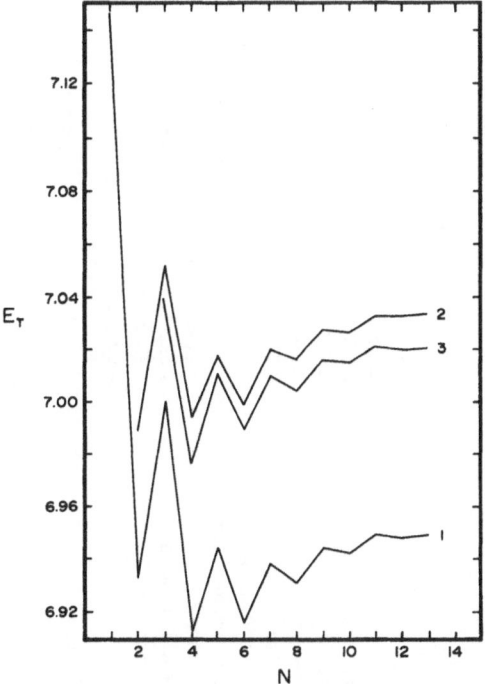

Fig. 6.1. Binding energy of the triton for the Reid soft core potential vs. number of terms N used in the expansion of the perturbation as a series of separable terms. The curves "1", "2" and "3" refer to three different choices of the unperturbed t-matrix: a single attractive term, and attractive and repulsive term, and two attractive and one repulsive term. From Afnan-Read (1973 B)

Table 6.2. Calculated Triton energy for Reid potential

Method	Lowest partial waves	All partial waves	Reference
Unitary pole approx.	-7.45	—	Bhatt et al. (1972) – See Section 5.5
"1A" approx.	-7.15	—	Afnan-Read (1973B) – See Section 5.6
"1A"; 13 terms pert.	-6.95	—	Afnan-Read (1973B) – Section 5.6, Fig. 6.1
"1A+1R"	-6.99	—	Afnan-Read (1973B) – Section 6.2, Fig. 6.1
"1A+1R", 13 terms pert.	-7.03*	—	Afnan-Read (1973B) – Section 6.2, Fig. 6.1
"2A+1R", 13 terms pert.	-7.02*	—	Afnan-Read (1973B) – Section 6.2, Fig. 6.1
Two-dim. Faddeev eqs. momentum space	-6.9*	—	Malfliet-Tjon (1972) – Kim (1973), Section 6.3
Faddeev eqs., co-ord sp.	-7.0*	-7.2*	Laverne et al. (1973) and priv. comm. Section 6.1
Variational, 2 core and 95 terms		$E_T < -6.8*$	Hennell-Delves (1973A) – Section 6.4, Fig. 6.3
Variational, 2 core extrap. to ∞ terms		-7.75 ± 0.5	Hennell-Delves (1972) – Section 6.4
Var, 484 oscillator terms	$E_T < -5.85$	$E_T < -6.3$	Jackson (1971) – Hadjimichael (1972)
Var, extrap. to ∞ oscillator terms	-6.25	-6.5	Jackson (1971) – Hadjimichael (1972) Section 6.4
Hyperspherical harmonics $k_{max} = 7$	$E_T < -6.258$		Bruinsma (1973) – Section 6.5
hh, extrap. to $k_m = \infty$	-6.8 ± 0.2*		Bruinsma (1973) – Section 6.5
hh, $k_m = 14$	$E_T < -6.62*$		Demin (1973) – Section 6.5
Brueckner-Bethe method		-7.207	Goldhammer (priv. comm.) Section 6.6
Experiment		-8.48	

The triton energy is given in MeV for calculations by methods discussed in Sections 5.5, 5.6, and 6.2 to 6.6. The lowest partial waves are ${}^1S$, and coupled ${}^3S_1 - {}^3D_1$. Starred results may be the most accurate.

use as the unperturbed Hamiltonian either the "single attractive" terms, "1 A + 1 R" or "2 A + 1 R", with unperturbed energy given as 5.32. The difference between the exact t-matrix for the Reid potential and that for the unperturbed Hamiltonian is expanded in a series of N separable terms. The rapid convergence with N is illustrated in Fig. 6.1. However, the value for E_T reached by increasing N is found to depend a little on the choice of the unperturbed Hamiltonian, varying from -6.95 MeV (for the choice "1 A") to -7.02 MeV (for the other two choices). The latter result of -7.02 MeV is included in Table 6.2, under "lowest partial waves".

Siebert et al. (1973) have recently applied t-matrix perturbation theory to find E_T for the Tabakin (1964) rank-two separable tensor potential, using the UPA for the unperturbed Hamiltonian. Equation (5.31) shows that the UPA misses the triton energy by $\frac{1}{3}$ MeV. Adding the first-order correction of 0.38 MeV to the UPA result of (5.31) we obtain

$$E_0 + \Delta E = -6.96 \text{ MeV} ,$$
$$E_T = -7.02 \text{ MeV} .$$

(6.2)

The latter result, copied from (5.31) is Brady's (1970) solution for Tabakin's potential. The agreement within 0.1 MeV is similar to Afnan-Read's result for the Reid potential, using their " 1A" as the unperturbed Hamiltonian.

6.3 Faddeev Equations for Local Potentials

We consider a central spin-independent local potential acting only in two-body S-states (Malfliet et al., 1969). Our result (5.17) above shows that the Faddeev equations in momentum space reduce, in this instance to a single homogeneous integral equation in two-dimensions. Osborn (1967) and Humbertson et al. (1968) pioneered in the difficult numerical problem of the solution of this two-dimensional integral equation, by replacement of the integral with a quadrature formula. Their methods and results were superseded by Malfliet and Tjon who write the Faddeev equations in inhomogeneous form, and look for the energy s for which the three-body T-matrix is singular. [In a similar way in the two-body problem, we could study either the equation $|B\rangle = \mathbf{G}_0(-B) v |B\rangle$ or look for a singular solution $\mathbf{t}(-B)$ of the LS equation, $\mathbf{t}(s) = v + v \, \mathbf{G}_0(s) \, \mathbf{t}(s)$.] Malfliet uses the following equation

$$\psi(p, q; s) = \phi(p, q; s) + (4\sqrt{3} q) \int_0^\infty q' \, dq' \int_L^U p' \, dp' \, t(p, p_1'; s - q^2)$$
$$\times \psi(p', q'; s) \, (s - q'^2 - p'^2)^{-1} .$$

(6.3)

Here $\psi(p, q; s)$ is the trinucleon pair wave function analogous to $\psi(p, q)$ in (5.17). [The momenta p are the same; Malfliet's q is a factor $\sqrt{\frac{3}{2}}$ times the momentum q defined in (5.1).] $\phi(p, q; s)$ is the momentum-space representation of $\mathbf{T}_1(s)|\psi\rangle$, where $\mathbf{T}_1(s)$ is defined in (5.7) as the solution of the LS equation for a two-body interaction between the second and third nucleons, using the three-body kinetic energy. The variable $p_1' = \sqrt{(p'^2 + q'^2 + q^2)}$. The upper and lower limits of integration are $L(q, q') = |2q - q'|/\sqrt{3}$ and $U(q, q') = (2q + q')/\sqrt{3}$. This type of equation has been solved in three different ways, by Malfliet et al., Kim (1969) and Haftel (1973). Malfliet uses a two-dimensional quadrature formula to replace (6.3) by a matrix-vector equation, which he solves by iteration. Kim uses "approximate product integration" in which he replaced the wave function by a finite sum of products of specified functions of p and q, with adjustable coefficient. (This amounts to use of the variational principle in momentum space.) Haftel solves by the method of Padé approximants (Tjon, 1970; Baker, 1965; Basdevant, 1972). That is, a function is represented as a polynomial of degree n_1 divided by a polynomial of degree n_2. The coefficients in the polynomials are adjusted to give agreement with (6.3). Any one of these methods involves difficult numerical work – clearly much more difficult than solution of the one-dimensional integral equation (5.22). When we worked with the separable approximation to the t-matrix, leading to (5.22), our concern was with the accuracy of the separable approximation. Now our concern must be with the accuracy of the numerical methods used to solve (6.3). For instance, Malfliet et al. estimates an accuracy of 0.1 MeV in the trinucleon energy calculated for the MTV potential: see Table 6.1. This result is for S-waves only. I also quote Malfliet's result (1969) including both S and D waves.

The reduction of the Faddeev equations to coupled two-dimensional integral equations for a local non-central force was performed by Ahmadzadeh et al. (1965), and Harper et al. (1970, 1972a). In the general case of interaction in the lower partial waves only (^1S and coupled $^3S_1 - {}^3D_1$) they find five coupled equations, of the general form (6.3). Malfliet et al. (1970) first approximated this system by three coupled equations, by neglect of the D-wave part of the wave function of the spectator nucleon. The solution of these three equations by the interative procedure mentioned above gave $E_T = -6.5$ MeV for the Reid soft core potential. Harper et al. (1972b) also use an iterative procedure to solve the five coupled equations for the Reid potential. They found $E_T = -6.7$ MeV. The most recent values by Kim et al. (private communication) and by Malfliet (1972) for the solution of four or five coupled two-dimensional integral equations are -6.8 and -6.9 MeV. See Kim (1974) for a detailed discussion of this work. I show in Table 6.2

the first Malfliet-Tjon value, for three integral equations, and the recent result of Kim et al. The remarks made above concerning solution of a single two-dimensional integral equation apply a fortiori in these cases: there is not complete agreement as to the numerical error involved in the determination of E_T.

Recently Gignoux and his coworkers have developed new methods of solution of the Faddeev equations for a local two-body potential. Again we start with a central spin-independent case, acting only in relative S-states. Benayoun et al. (1972) expand the Green's function $\mathbf{G}_0^{(3)}(E)$ in the Faddeev equation (5.12), giving a coupled set of one-dimensional integral equations. Alternatively, they combine (5.4) and (5.5) in the form

$$(E_T - \mathbf{H}_0^{(3)} - V_\alpha)|\psi_\alpha\rangle = V_\alpha(|\psi_\beta\rangle + |\psi_\gamma\rangle).\tag{6.4}$$

This equation can be expressed either in the momentum or in the configuration representation or in a mixed representation: configuration space for the left side, and momentum space for the right side giving a integro-differential equation. Gignoux et al. (1972) and Laverne et al. (1973) use the configuration representation on both sides. This gives another integro-differential equation, namely the Noyes equation (1970).

$$[\partial^2/\partial x^2 + \partial^2/\partial y^2 + E_T - V(x)]\,\psi(x, y)$$

$$= V(x)\,xy \int_{-1}^{1} x'^{-1} y'^{-1} \psi(x', y')\,\mathrm{d}u,\tag{6.5}$$

$$2x' = (x^2 - 2\sqrt{3}\,xyu + 3y^2)^{\frac{1}{2}},$$

$$2y' = (y^2 + 2\sqrt{3}\,xyu + 3x^2)^{\frac{1}{2}}.$$

Here x is the separation of a pair of nucleons, and y is proportional to the distance of the spectator nucleon from the center of mass of the "interacting pair": cf., (5.1). $\psi(x, y)$ is the eigenfunction. The kinetic energy operator $\mathbf{H}_0^{(3)}$ in (6.4) gives the second partial derivatives in (6.5). The changes of variables involved in expressing $|\psi_\beta\rangle$ and $|\psi_\gamma\rangle$ in terms of x and y gives the complicated form inside the integral on the right side of (6.5): cf., the analogous treatment in the momentum representation (5.15). The variable u in the integral is the cosine of the angle between the vectors x and y.

Gignoux (1972) and Laverne (1973) have generalized this derivation to give the five coupled integro-differential equations for a tensor force, neglecting the interaction in higher partial waves. They solve the coupled equations using a mesh in polar coordinates to evaluate both the partial derivatives on the left and the integrals on the right. They apply a scheme of "inverse iteration" for a guessed value of the eigenvalue E_T. Their result $E_T = -7.0$ for the Reid potential is given in

Table 6.2. They also calculate the effect of higher partial waves for this potential using Rayleigh-Schrödinger perturbation theory, giving $E_T = -7.3$ MeV for all partial waves. [Laverne (private communication) has revised this number to read $E_T = -7.2$ MeV.]

6.4 Variational Expansions in Ordinary Space

The variational method of minimizing the expectation value of the Hamiltonian for assumed normalized trial functions in ordinary space, with adjustable parameters, is as old as all other methods of calculation of the trinucleon energy put together. For a central spin-independent local potential, it gave very useful results almost forty years ago: Thomas (1935) showed that the triton energy decreased without limit as the range of the nucleon-nucleon force approached zero. The value of E_T is sensitive to the range of the force for a finite range; thus Thomas was able to give the first determination of the range of the force.

I illustrate Thomas' calculation for the case of a central spin independent local Yukawa potential, treated by Blatt et al. (1952).

$$V(r) = -V_0 \exp(-r/\beta)/(r/\beta),$$
$$V_0 = s'/0.59531 \, \beta^2, \tag{6.6}$$
$$\beta = b/2.1196.$$

Here s' is the strength parameter and b is the intrinsic range. Blatt uses a trial function with a single adjustable parameter, K.

$$\phi_K = \sqrt{4/7} \, K \exp[-\tfrac{1}{2} K(r_{12} + r_{13} + r_{23})]. \tag{6.7}$$

The expectation value $\langle H \rangle$ of the Hamiltonian is calculated analytically as a function of K, and K chosen to minimize $\langle H \rangle$. The result is

$$\langle H \rangle = -B(s')/b^2 \tag{6.8}$$

where $B(s')$ is positive for $s' \gtrsim 0.90$. The "collapse" as b goes to zero is evident. If we fix $s' = 1.139$, for instance, then Blatt gives us the result, $\langle H \rangle \simeq -2/(b^2)$. Setting $\langle H \rangle = ME_T/\hbar^2$ gives an intrinsic range of about 3 fm, which is not unreasonable. (Note the Blatt value of s' is an average of triplet and singlet strengths. The intrinsic range b is nearly equal to the effective range r_0 for the singlet potential, but b is about twice the effective range for the triplet case.)

Erens et al. (1971) have made a careful variational calculation for the MT(V) potential. Their result $E_T = -7.778$ MeV is given in Table 6.1. Note that with a variational calculation they automatically get an upper limit for E_T for all two-body partial waves. Erens' use of four significant figures shows his confidence that he has chosen a good trial function.

The basic problem in any variational calculation is the choice of a good enough trial function. The simple choice (6.7) is clearly inadequate when we use a two-body potential with a strong short-range repulsion, and with strong non-central forces: i.e., a realistic potential such as that of Hamada-Johnston, or Reid. Delves et al. (1969) and Delves (1972) discuss these problems carefully in their review articles.

Delves (1972) shows that the variational method gives very accurate results for a local central potential: e.g., a Yukawa potential discussed above, or an exponential potential, with a hard core. In either case, the equivalent two-body method (ETBM) provides sufficient accuracy. The ETBM corresponds to the Hartree approximation in atomic physics: the trial function is assumed to be a product of functions each involving only two nucleons. The function is chosen to minimize the triton energy. (Note that 6.7 is a special case of the ETBM, since we have arbitrarily chosen the form of the function, except for the choice of the parameter K. See Delves for discussion of the accuracy of 6.7.)

For an exponential with a hard core, the ETBM also is very accurate, giving the triton energy to with 0.03 MeV. Delves treats a two-body local Yukawa potential, using a "core function" of the form 6.7 together with an expansion of the form

$$\psi_N = \sum_{i=1}^{N} a_i S[\exp(-\alpha(jr_1 + kr_2 + lr_3))] . \tag{6.9}$$

As Delves points out, it is much easier to adjust the linear parameters a_i in (6.9) to minimize the triton energy, than to adjust the parameter K in (6.7) since the energy does not depend linearly on K. (S is a symmetrization operator; j, k, and l are integer dependent on i.) Nevertheless the use of one or two core terms is very useful, as is shown in Table 6.3, taken from Delves. A single core term gives an energy of

Table 6.3. Variational calculation of triton energy

Number of terms	With core	Without core
1	− 50.35	− 43.30
5	− 50.81	− 49.76
10	− 50.85	− 50.51
14	− 50.90	− 50.76
Exact	− 50.90	

Energy in MeV, for a local spin-independent Yukawa potential. The trial function without core uses form 6.9; the core is of form 6.7. From Delves (1972).

−50.35 MeV in the example treated, nearly as low as the energy using 10 terms in the expansion (6.9), without a core.

When a non-central potential is used, the trial function must include other triton states besides the completely symmetric S state considered above: namely the mixed symmetry S' state, and the principal D state (Delves et al., 1969). Each state in the trial function is chosen with core terms, and an expansion similar to (6.9). For the Hamada-Johnston potential, Delves and Hennell (1971) use an expansion with no core term, and with one, two or three core terms. The integer Q shows the number of adjustable parameters: Q of 7 corresponds to 93 parameters. As for the central potential with results in Table 6.3, the use of the core term greatly improves convergence, though in this more complicated case a Q of at least 3 is needed to get within an MeV of the triton energy. The use of a third core term does not help appreciably.

I shall not attempt here a discussion of the numerical methods involved in computation of the triton energy with a trial function, or of the extrapolation to an infinite number of adjustable parameters. The upper limits are clear: e.g., from Fig. 6.2 the energy for the Hamada-Johnson potential is less than −6.5 MeV. The error in the extrapolated energy $E_T = -6.5 \pm 0.2$ MeV involves both the numerical errors in evaluation of the expectation value of the Hamiltonian with a complex trial function, and also the errors of extrapolation.

The success of a variational calculation of this type clearly depends on clever choices of the core function, and of the set of functions used for the expansion. (One might be clever enough to guess a single term core function that was a good enough approximation to the true wave function, or one might be unfortunate enough to use a trial function orthogonal to the true wave function.) It is clear that Delves et al. are able to make clever enough choices for a potential with a hard core. This cleverness is based on long experience with a given form of potential. The rapidity of convergence is at present not as rapid for the Reid soft core potential. I show in Fig. 6.3 results by Hennell and Delves presented in 1971, but published in 1973. As in Fig. 6.2, use of a core helps considerably; but the convergence is poorer for the Reid potential than that shown for the Hamada-Johnston potential. Figure 6.3 gives an upper limit for the triton energy of −6.8 MeV. Hennell and Delves (1972) extrapolate to an infinite number of parameters in their trial function, and find $E_T = -7.75 \pm 0.5$ MeV. Both results are given in Table 6.2.

A less subjective but apparently less successful choice of trial functions is made in the variational calculations for non-central forces by Jackson et al. (1970, 1971) and Hadjimichael and Jackson (1972). They choose a complete set of functions, and systematically include

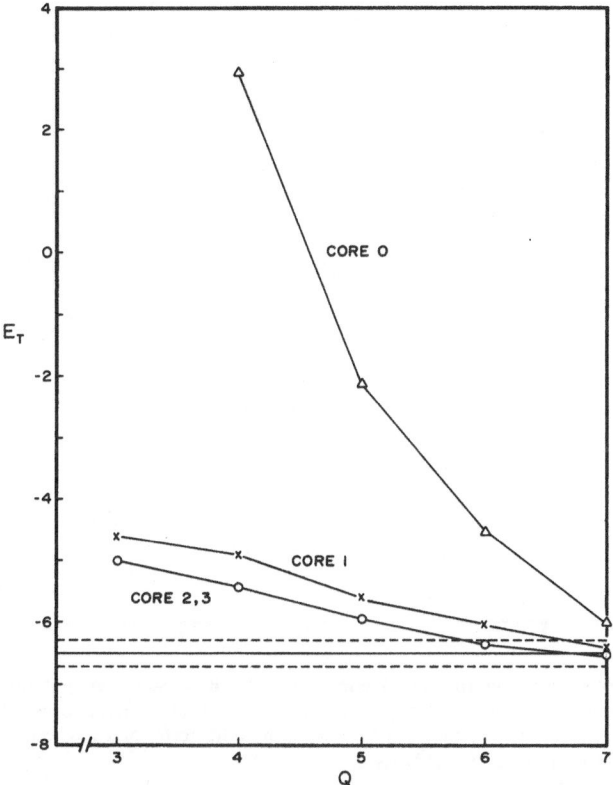

Fig. 6.2. Upper bounds on the triton energy for the Hamada-Johnston potential vs. Q, which corresponds to the number of parameters in the variational expansion. The three curves are for no "core term" in the trial function, one core term, and two or three core terms. The converged value (solid and dotted horizontal lines) shows $E_T = -6.5 \pm 0.2$ MeV. From Delves (1972)

more and more of them in their variational calculation. (Their use of solutions for a harmonic oscillator is the same as that used in *some* nuclear Hartree-Fock calculations; so the speed of convergence is of great interest to the groups performing this type of HF calculations for larger nuclei.) It seems likely that this systematic method would not converge as rapidly as the "method of clever choice"; but perhaps with modern computers speed of convergence is of less importance than it used to be. I show four results of Jackson's work in Table 6.2: use of 484 terms in the trial function, and extrapolated to an infinite number of terms, both without and with higher partial waves in the Reid soft core potential. Note the disagreement of about an MeV between Hennell's $E_T = -7.75 \pm 0.5$ MeV and Jackson's $E_T = -6.5$ MeV (both

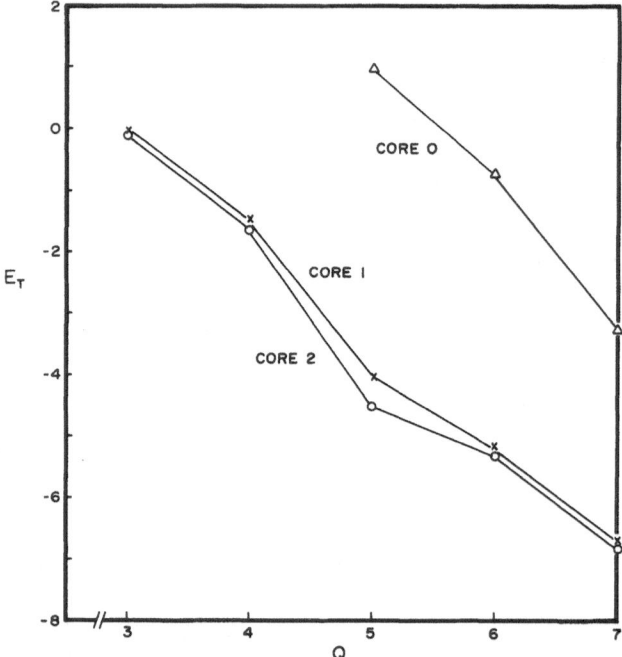

Fig. 6.3. Upper bounds on the triton energy for the Reid soft core potential, vs. Q, which corresponds to the number of parameters in the variational expansion. The three curves use no "core term" in the trial function, one core term, and two core terms respectively. From Hennell-Delves (1973 A)

extrapolated to an infinite number of terms, and both including all partial waves in the Reid potential). Strayer and Sauer are extending this expansion to still more terms; I await their results with interest.

It would certainly be helpful to have good lower bounds for E_T. Two methods have been used. Delves et al. (1969) and Delves (1972) discuss the comparison of the expectation value of the Hamiltonian $\mathbf{H}$ and of $\mathbf{H}^2$, as a way of finding the lower bound. Since $\langle H^2 \rangle$ is very different from $\langle H \rangle^2$ the lower bound is far from the upper bound. An alternative method was developed by Hall (1968) in which the energy E_T of the three-nucleon system for a central spin-independent potential is shown to obey the inequality

$$E_T > -2B'. \tag{6.10}$$

Here B' is the "two-body binding energy" for the solution of the two-body problem with the usual kinetic energy operator, and with the "two-body potential" equal to $1\frac{1}{2}$ times the actual two-body potential.

[Equation (6.10) is derived by straightforward manipulation of the three-body Schrödinger equation (neglecting possible three-body forces), for a local or non-local potential. Hall generalized it to an average of the spin singlet and spin triplet values of B' if there is spin-dependence.]

The simple Hall-Post lower bound turns out to be of some use in cases where it is easy to solve accurately for E_T, such as a central spin-independent potential without short-range repulsion: the lower bound (6.10) is of order an MeV below the true eigenvalue (Humbertson et al., 1968). But when appreciable spin-dependence, or short range repulsion, or both are introduced, the Hall-Post lower bound falls an order of magnitude or more below calculated values of E_T (Brady et al., 1971). Thus for the Reid potential, the lower bound is -270 MeV. Neither the use of $\langle H^2 \rangle$ nor the use of the Hall bound provides useful information for this case.

Recently Nissimov (1973) has developed a method which, at least in a sample two-body problem, gives much higher lower bounds. Application to the three-body problem will be of great interest.

In the absence of useful lower bounds, the only test of convergence of a variational expansion is that it *appears* to have converged: e.g., Fig. 6.2 for the Hamada-Johnston potential. Of course, an upper bound calculation is very useful if it is below the experimental value; but if it falls above experiment, it is hard to draw rigorous conclusions. Still, one can argue (Delves, 1972) that a firm upper bound is better than a method which might have large numerical errors of either sign, such as the solution of coupled two-dimensional integral equations. The most reliable test of the reliability of several methods seems to be their agreement on the calculated triton energy within the claimed uncertainties in each calculation.

I now turn to another variational method. Logically speaking, it does not deserve a new section; but I give it one since it is an expansion in an unfamiliar complete set of functions: "hyperspherical harmonics".

6.5 Variational Expansion in Hyperspherical Harmonics

This method was pioneered by Delves (1959, 1960) and by Simonov (1966). The method was improved by Fabre (1969, 1971). It was recently reviewed by Simonov (1972) and Louck et al. (1972). Efros (1972) applied the method to two-body tensor forces, and Demin et al. (1973) and Bruinsma et al. (1973) have recently given results for E_T for the Reid potential.

The basic idea is to generalize the spherical harmonics used in the two-body problem: hence the name hyperspherical harmonics, or h.h.

Thus in the two-body problem we express the wavefunction in terms of the vector r_{12} joining the two particles, and separate the three-dimensional partial differential equation $H|\psi\rangle = E|\psi\rangle$ into a one-dimensional equation for the magnitude r_{12} and another equation for the angular dependence. [Note $r_{12}\psi(r_{12}) = u_l(r_{12}) Y_{l,m}(\omega)$, where we combine the usual angles θ and ϕ for r_{12} into a single letter ω.] Primes denote derivatives with respect to r_{12}. The radial function $u_l(r_{12})$ obeys

$$-u_l'' + [l(l+1)/r_{12}^2 + V(r)]\, u_l(r_{12}) = E_l u_l(r_{12}),$$
$$l^2 Y_{l,m}(\omega) = l(l+1)\, Y_{l,m}(\omega).$$

$$(6.11)$$

where $-l^2/r_{12}^2$ is the part of the kinetic energy operator that depends on ω. The $Y_{l,m}(\omega)$ are the ordinary spherical harmonics, eigenfunctions for the squared angular momentum l^2 and also for l_z. The value of the non-negative integral quantum number l affects the radial function $R_l(r_{12})$ and the energy E_l through the centrifugal term which in our units is $l(l+1)/r_{12}^2$.

Of course we obtain uncoupled equations (6.11) for the radial function $u_l(r_{12})$ only if we have a central potential $V(r_{12})$; in general we obtain N coupled differential equations such as the coupled equation for $l=0$ and 2 for the deuteron ground state. The limitation to two values of orbital angular momentum is the result of two quantum rules. First, the total angular momentum j is the vector sum of l and the spin angular momentum S. From experiments on the deuteron, j and S are each 1: this limits l to the values 0, 1, and 2. But parity is a good quantum number (to an accuracy of order one part per million), and the deuteron parity is even: then l is restricted to the values 0 and 2.

For the three-body system with equal masses, we use Jacobi coordinates similar to those of (5.1) and Fig. 5.1. We define

$$\xi_1 = r_2 - r_1,$$
$$\xi_2 = \sqrt{3}(r_3 - R).$$

$$(6.12)$$

Here r_1, r_2, r_3, and R are the coordinates of the three particles, and of the center of mass respectively. The two vectors ξ_1 and ξ_2 are combined to give a single vector r in 6-dimensional space, which has a hyper-length r, and 5 angles, which we shall denote by the letter Ω. The length r is

$$r = \sqrt{\xi_1^2 + \xi_2^2}.$$

$$(6.13)$$

The kinetic energy part of the three-body Hamiltonian, $H_0^{(3)}$ is now written as the sum of an operator involving r and one involving Ω. For a single "partial wave" with quantum number L (called the "grand orbital"),

$$\mathbf{H}_0^{(3)}\psi_L(r) = -\psi_L''(r) - (D-1)\,\psi_L'(r)/r + L(L+D-2)\,\psi_L(r)/r^2\,. \qquad (6.14)$$

Here $D=6$ for a three-particle system. In general $D=3(A-1)$ and gives the dimensionality of 6 for the hyperspace used for $A=3$. For $A=2$, $D=3$ and (6.14) agrees with the familiar expression for the two-body kinetic energy. We separate into partial waves, expanding $\psi(r)$

$$\psi(r) = \sum_{L=0}^{\infty} Y_{(L)}(\Omega)\,\psi_{(L)}(r)\,. \qquad (6.15)$$

Here (L) denotes 5 quantum numbers (replacing the two l and m for the two-particle system); and $Y_{(L)}(\Omega)$ is a hyperspherical harmonic of the 5 angles. These hyperspherical harmonics are eigenfunctions of the operator $\mathbf{L}^2$, which includes the part of the kinetic energy that involves derivatives with respect to angles. They were first introduced by Appel (1922). Simonov (1966) chose a linear combination of h.h., corresponding to a definite value of the total orbital angular momentum of the system: this combination is called K-harmonics.

Except for the special case of a two-body harmonic oscillator potential, the sum of the three two-body potentials is "non-central" in terms of the hyperspherical harmonics: that is states of different L are coupled by the potential energy. Simonov (1966) and Fabre (1969) give formulas for the matrix elements $V_{LL'}$, between states with quantum numbers L and L'. Using these matrix elements, together with (6.14) for the kinetic energy, we obtain an *infinite* set of coupled differential equations in the single hyperspherical radial variable r. The crucial question is whether, for some given choice of the two-body potential, we are able to truncate this set at a not unreasonable value, so that we can solve for the eigenvalue and eigenfunction.

Following Fabre (1969), define a radial kinetic energy operator

$$\mathbf{T}_{2k} = -(\hbar^2/m)\left[d^2/dr^2 + 5r^{-1}\,d/dr - 4K(K+2)/r^2\right]\,. \qquad (6.16)$$

The first three coupled equations for a central spin-independent two-body force are

(a) $(\mathbf{T}_0 - E)\,\psi_0 + 12\pi^{-\frac{1}{2}}\{V_0\psi_0 + \sqrt{3}\,V_4\psi_4 - \sqrt{8}\,V_6\psi_6\} = 0\,,$

(b) $(\mathbf{T}_4 - E)\,\psi_4 + 12\pi^{-\frac{1}{2}}\{V_0\psi_4 + \sqrt{3}\,V_4\psi_0 - \tfrac{6}{5}\sqrt{6}\,V_6\psi_6\} = 0\,, \qquad (6.17)$

(c) $(\mathbf{T}_6 - E)\,\psi_6 + 12\pi^{-\frac{1}{2}}\{(V_0 + \tfrac{9}{5}V_4)\,\psi_6 - \sqrt{8}\,V_6\psi_0 - \tfrac{6}{5}\sqrt{6}\,V_6\psi_4\} = 0\,.$

The multipole $V_{2K}(r)$ is expressed in terms of the two-body potential. Define

$$v(k_{12}) = (2\pi)^{-3}\int v(r_{12})\exp(i\boldsymbol{k}_{12}\cdot\boldsymbol{r}_{12})\,d^3r_{12}\,. \qquad (6.18)$$

Then

$$V_{2k}(r) = (4\pi)^{\frac{3}{2}} r^{-2} \int\limits_0^\infty v(k)\, J_{2k+2}(kr)\, \mathrm{d}k\,. \tag{6.19}$$

For a two-body gaussian (or sum of gaussians) the multipole $V_{2k}(r)$ is a modified Bessel function.

Fabre (1969) made a significant improvement to this expansion, by introducing a linear combination of Simonov's hyperspherical harmonics, which Fabre calls the "optimal subset". (They are also called "potential harmonics".) The subset is chosen so that a given type of potential will have all its non-zero matrix elements between members of the subset. This choice, subsequently adopted by Erens et al. (1971) for central potentials and by Fabre (1971), and Efros (1972) for tensor potentials, gives a substantial improvement in the speed of convergence: e.g., in the central case, 27 members of Fabre's subset take into account 75 members of Simonov's original set.

There are two reasons for hoping for rapid convergence of the series typified by (6.17). First, we need only even values of L, for the ground state with even parity, since the wavefunction has parity $(-1)^L$. We therefore use the integer $k = L/2$. Second, the centrifugal barrier in (6.14), namely $L(L+4)/r^2$, increases rapidly with increasing L, and should decrease the effect of the higher hyperspherical harmonics. Since the expansion (6.15) is an example of a trial function, the value of E is a monotonic decreasing function of the maximum value chosen for k and is an upper bound.

I illustrate the speed of convergence for the Malfliet-Tjon potential V. Results by Erens et al. (1971) using Fabre's optimal subset are shown in Figs. 6.4 and 6.5 vs. the largest value of k, called k_m. We see that the first few harmonics give drastic changes in the triton energy; but that convergence to better than 0.1 MeV seems to be achieved by the use of about 20 coupled differential equations. (Note the expanded scale in Fig. 6.5.) The good agreement with other results for this potential shown in Table 6.1 strengthens the "internal evidence" that convergence has been achieved. [Two parenthetical remarks. First, the convergence is slower for MT-V than for the other seven potentials used by Erens, due to the strong short range repulsion in this potential. Second, Erens solves the k_m coupled differential equations by introducing trial functions for each of the $\psi_L(r)$ in (6.15) and adjusts parameters in these functions to minimize the energy; while Beiner et al. (1971, 1972) solve the equations directly. Either method should be satisfactory.]

Once we introduce non-central (or central spin-dependent) two-body forces, the number of coupled equations for a given value of k_m increases greatly. Since machine capabilities limit the number of

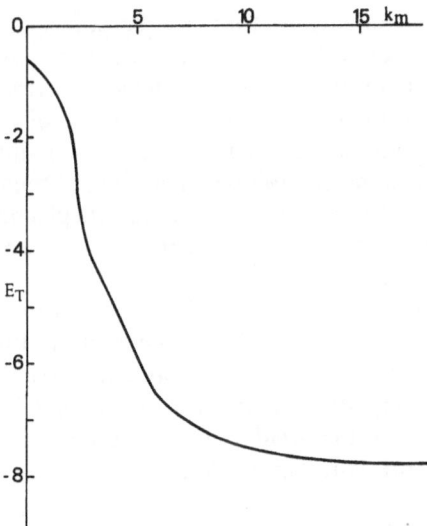

Fig. 6.4. Convergence of expansion in hyperspherical harmonics for the MT-V central potential. Triton energy E_T in MeV vs. the number of coupled differential equations, k_m. See Erens et al. (1971)

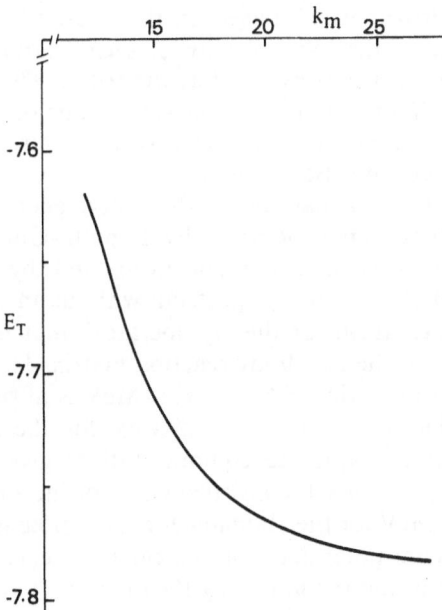

Fig. 6.5. Convergence of expansion in hyperspherical harmonics for the MT-V potential: a replot of Fig. 6.4 on an expanded scale

coupled differential equations that can be solved, this gives a severe limitation to the value of k_m we can use. It is now particularly useful to use Fabre's (1971) optimal subset for tensor forces, since this allows reaching a higher value of k_m for a fixed number of coupled differential equations. For example, Bruinsma et al. (1973) did *not* use Fabre's optimal subset; machine capabilities then limited them to $k_m = 7$, giving $E_T < -6.258$ MeV for the Reid potential (neglecting higher partial waves). They extrapolate to infinite k_m, and obtain E_T of -6.8 ± 0.2 MeV. Both values are given in Table 6.2.

The one calculation I have seen using the optimal subset for tensor forces is that of Demin et al. (1973). They show that convergence is achieved for $k_m = 14$, if the two-body potential is that of Eikemeier-Hockenbroich (1971) or Gogny-Pires-de Tourreil (1970). The one number they give for the Reid potential is also for $k_m = 14$: namely, $E_T < -6.62$ MeV, presented in Table 6.2.

6.6 Brueckner-Bethe Method

The above method was designed to work for nuclear matter, and subsequently for finite nuclei. The latter are generally treated in a "local density approximation" based on the two-body reaction matrix for nuclear matter at the local density. It seems surprising that these methods work for a nucleus as small as the triton. The results of these calculations are of interest both as regards the value of the triton energy for a specified potential, but particularly as regards a test of the accuracy of the Brueckner-Bethe method.

Two independent calculations of the triton energy have recently been made using Brueckner methods: by Tripathi-Goldhammer (1972) and Goldhammer (private communication), and by Sotona (1973). Tripathi uses the Bethe reference spectrum without an energy shift, and uses an approximate form for the "Q operator" that takes account of the Pauli principle in the two-body reaction matrix. Their result for the Hamada-Johnston potential of $E_T = -6.24$ MeV is in surprisingly good agreement with Delves' result of -6.5 MeV for the same two-body potential. Goldhammer (private communication) gives results with a more careful treatment of self-consistency and of the Pauli operator Q: he quotes -6.481 MeV for the Hamada-Johnston potential, and -7.207 for the Reid soft core potential. Both calculations agree very well with other calculations of the triton energy for the same potentials.

Sotona makes similar calculations, using a shifted reference spectrum. He calculates for rank-two separable potentials: Tabakin (1964) and Mongan (1969) among others. His results for E_T vary with the oscillator frequency chosen for the shell model base, but tend to be within $\frac{1}{4}$ or $\frac{1}{2}$ MeV of values of E_T from solution of the Faddeev equations (Brady, 1970).

It is not completely clear how these workers treat the center of mass motion, which usually poses a serious problem in applying the Brueckner-Bethe method to very light nuclei. Their results are certainly encouraging; even though one must admit that there are adjustable parameters in their calculational formalism.

6.7 Triton Energy for Reid Potential

Table 6.1 illustrates the good agreement among a variety of methods for the triton energy calculated with a central spin-independent potential with a strong repulsion. The excellent agreement between the last two rows (variational and hyperspherical harmonics) is particularly striking.

The situation is slightly more confused in the comparison of 19 different results for the triton energy for the Reid soft core potential, shown in Table 6.2, along with the experimental value of E_T. I shall try to pick the reliable methods, even though the results quoted are quite recent; and are in the process of being checked or refined, or both. One particularly striking disagreement is between the variational result of Hennell and that of Jackson. I believe that Jackson's extrapolation to infinity is inaccurate. Other disagreements can be interpreted as i) due to the approximate nature of the calculation (e.g., the unitary pole *approximation*), or ii) due to the possible important contributions from higher partial waves (say Harper's result as compared with Hennell's), or iii) due to truncation, as in Bruinsma's value for $k_m = 7$. I find it gratifying that so many methods agree quite well.

Consider the six starred results in Table 6.2 for triton energy, treating only the lowest partial waves (^1S, and coupled $^3S_1 - ^3D_1$). Each is "internally consistent": e.g., the Afnan-Read perturbation results illustrated in Fig. 6.1. They also each agree with the others, or with an average, $E_T = -6.9 \pm 0.1$ MeV. (I interpret Demin's upper bound as showing that an extra $-\frac{1}{4}$ or $-\frac{1}{3}$ MeV of triton energy is contributed by higher hyperspherical harmonics.)

There are fewer results in the column for all partial waves, and I star only two of these: Laverne's value, using Rayleigh-Schrödinger perturbation theory to find the contribution of the higher partial waves to the triton energy, and Hennell-Delves upper limit. (The quoted error in the Hennell-Delves extrapolated value is uncomfortably large; and I ignore the Jackson extrapolated result.) I assume Laverne's value of -0.2 MeV for ΔE_T for higher partial waves and adopt $E_T = -7.1$ MeV. I find it hard to estimate the uncertainty.

There is a clear disagreement between the calculated -7.1 MeV and the experimental triton energy of -8.48 MeV. I shall return to discussion of this disagreement in Section 6.9, after discussion of the coulomb energy of ^{3}He.

6.8 Coulomb Energy Difference

The calculated coulomb energy difference between ^{3}H and ^{3}He used to be regarded as a check on the validity of the trinucleon wave function used (Blatt et al., 1952). Charge symmetry of nuclear forces was assumed to hold. But for some time Okamoto (1964, 1971) has argued that the disagreement between the calculated coulomb energy and the experimental energy difference of 0.76 MeV should be interpreted as a failure of the assumed charge symmetry of nuclear forces. I discussed possible sources of deviations from charge symmetry in Chapter II. Fortunately we now have a (almost) model independent method of calculating the coulomb energy, due to Fabre (1972) using the charge form factors of ^{3}H $- ^3$He determined from experiments on elastic electron scattering from ^{3}H and ^{3}He.

It seems off-hand that there should be a relation between coulomb energy and the charge form factor since both depend on the distribution of proton density in the trinucleon. But there is a difficulty since the coulomb energy of ^{3}He in first order perturbation theory depends on ξ_1, of Eq. (6.12), if we suppose that the first and second nucleons are protons. On the other hand, the electron scattering depends on the distance of the protons from the center of mass, which is proportional to ξ_2. However, Fabre points out that if the hyperspherical quantum number k is even, it turns out that the trinucleon wave function is unchanged when ξ_1 and ξ_2 are interchanged. In the oversimplified approximation of treating both proton and neutron as point charges and of using the same wavefunction for ^{3}H and ^{3}He, and therefore the same form factor $F(k)$ for electron scattering, the coulomb energy of ^{3}He has the simple form

$$E_c = (4e^2/\pi\sqrt{3}) \int_0^\infty F(k)\,dk. \qquad (6.20)$$

This expression can be understood as the expectation value of the coulomb interaction e^2/k^2 evaluated using the charge density distribution $F(k)$ in momentum space. [The factor outside the integral comes from Fourier transforms, and from the $\sqrt{3}$ in definition (6.12) for ξ_2.]

The complete expression of Fabre's is more complicated for three reasons: i) the proton is not a point charge, but instead has an electric form factor $G_E(q)$ where q is the momentum transfer; ii) the neutron also has a distribution of charge; iii) the mixed symmetry state ($k = 1$) gives different electric form factors $F_H(q)$ and $F_{He}(q)$ for ^{3}H and ^{3}He. (Note that we assume that isospin is a good quantum number.) Including the first effect changes the coulomb interaction, by introducing a factor $G_E^2(k/\sqrt{3})$, and also changes the trinucleon squared wave function from $F(k)$ to $F(k)/G_E(k)$. Equation (6.20) is replaced by

$$E_c = (4e^2/\pi \sqrt{3}) \int_0^\infty dk\, G_E^2(k/\sqrt{3})\, F(k)/G_E(k) \,. \tag{6.21}$$

When we include the second effect listed above, we must complicate our notation to allow for two different electric nucleon form factors: the isovector $G_{EV}(k)$ and the isoscalar $G_{ES}(k)$. Fabre treats the third effect by using the measured form factors F_H and F_{He} to take account of mixed symmetry state. His complete formula reads

$$E_c = (2e^2/\pi \sqrt{3}) \int_0^\infty 4G_{ES}(k/\sqrt{3})\, \{[2F_{He}(k) + F_H(k)]/(6G_{ES}(k))$$
$$- [1/3\, G_{EV}(k)]\, [F_H(k) - F_{He}(k) - \tfrac{1}{4}(G_{ES}(k) - \dot{G}_{EV}(k))] \tag{6.22}$$
$$\cdot [2F_{He}(k) + F_H(k)]/G_{ES}(k)\} = 0.65 \text{ MeV} \,.$$

It includes states $k = 0$, 1, 2, 4, 6, ... The numerical value of 0.65 MeV from *experimental* quantities on the right side of (6.22) needs small corrections for values of $k \geq 3$ and for the effect of $n - p$ mass difference on the kinetic energy term in the Schrödinger equation. Fabre concludes

$$0.64 < E_c < 0.68 \text{ MeV} \,. \tag{6.23}$$

Comparison with experiment shows the contribution of some 0.1 MeV due to failure of charge symmetry of nuclear forces.

Fabre points out that a variety of calculations that give reasonable agreement with trinucleon form factors [at modest values of momentum transfer which give significant contributions to the integral (6.22)] must, and indeed do, give results for E_c close to those quoted above.

Thus we use the energy splitting of the trinucleon isodoublet to learn about the nuclear forces, rather than about the accuracy of one or another solution of the trinucleon problem.

6.9 Comparison with Experimental Triton Energy

Let's return to Table 6.2 and the disagreement between the calculated energy $E_T = -7.1$ MeV (or thereabouts) for the Reid potential and the experimental value, $E_T = -8.48$ MeV. I choose 1.4 MeV. Why this disagreement of some 1.4 MeV? The reasons fall into 4 main groups: i) Calculational difficulties; ii) an incorrect choice of the two-nucleon potential (or corresponding off-shell t-matrix); iii) relativistic effects; iv) effects of three-body forces.

As remarked above, considerable progress has been made recently in new methods of calculation, so that calculational errors may not be much larger than 0.1 MeV, or much less than the other three sources of uncertainty in calculations of E_T. (I am discarding Jackson's value of -6.5 MeV to obtain the stated agreement with a calculated value of -7.1 MeV.)

I discussed the choice of a phenomenological nucleon-nucleon potential in Chapter II, where I emphasized the uncertainty in the mixture of central triplet and tensor forces, and a possible determination of p_D by measurements of deuteron polarization effects in elastic electron-deuteron scattering. The effect of this uncertainty on E_T was studied in Chapter V, using separable t-matrices with form factors of Yamaguchi shape. The results, Fig. 5.3, show that lowering the percentage D-state, p_D from 6.4% (for the Reid potential) to, say, 4.9% would lower E_T by $\frac{1}{2}$ MeV. Also, lowering the singlet effective range r_0 from 2.70 fm would lower E_T: but in fact the experimental r_0 is larger than 2.7 fm (Noyes, 1972 A). Similar conclusions have been reached by a comparison of calculations using local potentials. For instance, Laverne and Gignoux (1973) compare values of triton energy calculated for the de Tourreil-Sprung supersoft core with their values in Table 6.2 for the Reid soft core. The decrease of about $-\frac{1}{2}$ MeV from the Reid value is due to two changes in the potential: use of a lower percentage D state in the deuteron, and use of a supersoft core. The two effects of core shape and percentage D-state also have been studied by Hennell and Delves (1973 B), by variational calculations for the Hamada-Johnston hard core, and for the Gammel-Brueckner potential. The comparison of Reid soft core and Hamada-Johnston shows that the latter hard core increases the energy by about $\frac{1}{2}$ MeV. The decrease in D-state from 6.7 to 5% in changing from HJ to GB potentials lowers the energy by an MeV.

We can use Fabre's analysis above of the calculated coulomb energy of the trinucleon isodoublet to make a (small) correction for the assumption of charge symmetry in our calculation of E_T. In practice, the phenomenological nucleon-nucleon potential is fitted to proton-proton scattering and assumed to be the same for the neutron-neutron potential. But Fabre has shown that the neutron-neutron potential is some $\frac{1}{2}$% more attractive than the nuclear part of the proton-proton potential: this lowers the calculated triton energy of E_T by about 0.1 MeV. This correction is only some 7% of the discrepancy, but it helps.

The problem of the local vs. non-local character of the potential, or equivalently, of the correct prescription to extrapolate the t-matrix from on-shell to off-shell values, is a serious problem. It has been investigated for phase-equivalent potentials, which can give different deuteron wavefunctions; and also for potentials which are both phase equivalent and "eigenfunction equivalent". Each case can be investigated both for model central spin-independent potentials, and for a non-central potential, such as the Reid soft core.

The model case, allowing changes in the eigenfunction, was studied by Fiedeldey et al. (1969, 1972) using different rank-2 separable central spin-independent potentials adjusted to be phase equivalent to the potential of Tabakin (1965). The value of E_T varied from -8.23 MeV to -9.44 MeV, compared to -8.74 MeV for the original potential. This modest variation was for a special class of potentials obeying two criteria: i) there is a high overlap integral between the eigenfunction for Fiedeldey's rank-2 separable potential and that for the Tabakin (1965) potential; namely more than 0.9928 for the energies quoted above; ii) the rank-2 separable potential is not close to what Fiedeldey calls a "limiting case" for which the two-body bound state energy for his attractive term is very close to that for the original Tabakin potential. Afnan and Serduke (1973) also find only small changes in E_T if $|B\rangle$ is held fixed.

Recently Haftel (1973) assumed a local central spin-independent potential similar to MTV, and produced phase-equivalent potentials with a short-range unitary transformation. He found that E_T of -7.65 MeV could be varied from -7.0 MeV to -3.66 MeV. But if he confined himself to unitary transformations which preserved the eigenfunction, the variation was sharply restricted.

The more difficult calculational problem of finding the effects of off-shell behavior for a family of non-local potentials that are phase-equivalent to the Reid soft core (Haftel et al., 1971) was studied by Hadjimichael and Jackson (1972) and by Harper et al. (1972C). The singlet potentials have no constraint on the eigenfunction, since there is no two-body state. The triplet potentials are not eigenfunction equivalent, but they do have the constraint of giving (very nearly) the same deuteron quadrupole moment. Hadjimichael finds changes in E_T from -4.75 MeV to ca. -8.0 MeV (cf. the value -6.25 MeV in Table 6.2 for the Reid potential). On the other hand for the same 4 transformations Harper finds changes in E_T of only $\frac{1}{2}$ MeV. This numerical disagreement is unresolved at present: that is calculational difficulties have obscured the problem of the allowable variation of E_T with phase equivalent potentials.

Very recently Sauer and Tjon (1973) and Sauer and Stingl (1973) have used Sauer's phase equivalent ^1S t-matrices to find the change in triton energy. Sauer changes the off-shell t-matrix, by adjustments in the symmetric part of the half-off-shell t-matrix, and does not need explicit reference to any potential. The three examples treated by Sauer-Tjon changed the triton energy by 0.3 MeV, as compared to that for the Reid singlet t-matrix. Sauer-Stingl present a formalism to study changes in triton energy, but do not give numerical results.

In principle (though not yet in practice) we can insist on eigen-function-equivalent potentials for the spin-triplet case; but as Haftel (1973) emphasizes, we have the greater freedom of phase equivalence alone for the singlet potential. The poor knowledge of the off-shell t-matrix for the singlet case remains a major source of uncertainty in the trinucleon calculations. The close agreement we found between **t** and $\mathbf{t}^U$ for the Reid singlet (Table 3.3) is suggestive, but certainly isn't proof that phase-equivalent potentials should not give wildly different off-shell t-matrices. Sauer's examples give only small shifts in triton energy. The recent effective range expansion (Bahethi et al., 1972) for the off-shell t-matrix also suggests that we do not have tremendous freedom in the off-shell t-matrix. The problem of the freedom of phase equivalent t-matrices is not yet settled.

All calculations discussed above are non-relativistic. But a simple estimate (Blatt et al., 1952) shows that relativistic effects in the tri-nucleon should not be negligible:

$$v^2/c^2 \simeq 2\langle T\rangle/3\,c^2 \simeq 3\,\%. \tag{6.24}$$

Here v is a typical nucleon speed, and the expectation value of the tri-nucleon kinetic energy $\langle T\rangle$ is roughly 50 MeV.

We may then expect changes of order 3% in both the kinetic and potential energies in the trinucleon. But while we have a good theory for the relativistic corrections of about -1 MeV to the kinetic energy (Gupta et al., 1965), we do not have a relativistic theory for the t-matrix. (We don't even have a unique non-relativistic off-shell t-matrix!) From our present phenomenological base, relativistic effects on the t-matrix are closely connected to the other unsolved problem of extrapolation of the t-matrix to find off-shell values. Jackson and Tjon (1970) attacked these relativistic problems maintaining the assumption that the relativistic potential would be local, of the same form as Reid's but with modified values of its parameters, so that it would fit the deuteron and two-nucleon scattering in a theory with relativistic kinematics. They then solved the trinucleon problem, by an iterative solution of three coupled two-dimensional integral equations, with the modified Reid potential and relativistic kinematics. They compare with Malfliet's non-relativistic solution for the Reid potential, and find that relativistic effects lower E_T by about -0.25 MeV. (The value quoted is subject to error, due to inaccuracies in solving each trinucleon problem.)

The fourth, and in many ways the most interesting, effect listed above is that of the three-nucleon force. I discussed it qualitatively in Chapter I, and shall not go much farther here. Physicists calculate the contribution of one or another diagram that involves all three nucleons: e.g., that of Fig. 6.6 for Fujita et al. (1957) and Coury et al. (1963).

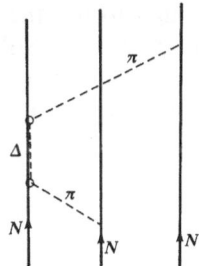

Fig. 6.6. A diagram for a three-nucleon potential, involving exchange of two pions, with the solid and dotted line for the intermediate state drawn on the left showing the Δ isobar. From Fujita et al. (1957) and Coury et al. (1963)

Pask (1967) evaluated the expectation value of this potential using Delves' variational wave function for the Hamada-Johnston potential. He finds that E_T is lowered by 1.4 MeV. (See Phillips, 1966; Laroze, 1970; and Delves et al. (1971) for other results with phenomenological three-nucleon forces.)

G. E. Brown (1972) gives a summary of recent work on the three-nucleon force. He suggests the likely importance of other diagrams, besides that of Fig. 6.6. Brown expects that the force due to other diagrams will lower the trinucleon energy an amount comparable to the 1.4 MeV found by Pask.

It is clear that there are plenty of ways of explaining the discrepancy of some -1.4 MeV between the value of E_T calculated with the Reid soft core potential, and the experimental value. The failure of charge symmetry contributes -0.1 MeV. Change of the choice of a local potential could contribute $-\frac{1}{2}$ MeV. A different off-shell extrapolation could contribute $\pm\frac{1}{2}$ MeV. (This numerical value is controversial, due to the disagreement between Hadjimichael et al. and Harper et al.) Relativistic effects contribute about $-\frac{1}{4}$ MeV. Three-body forces contribute the order of -1.4 MeV. We would easily find that the total correction terms are twice as large as we need to obtain agreement with experiment.

Some of the future work needed on these correction terms is straightforward but difficult: e.g., a better determination of the deuteron percentage D-state, or a better calculation of relativistic effects on the triton energy. But the tangled problem of extrapolation of the two-body t-matrix to off-shell values and the phenomenological three-body force we should assume is still facing us: we have not untied this portion of the Gordian knot. The numbers quoted above for the effects of phase-equivalent interactions do give us the *hope* that the uncertainties in the two-body off-shell t-matrix for "reasonable extra-polations" may be small enough to use the triton energy to determine the contribution of three-body forces. Still, it is always difficult to obtain firm conclusions with comparison with a *single* experimental number: comparison with an experimental *function* would likely give us more information.

6.10 Trinucleon form Factor

We therefore turn to the comparison of experimental and calculated values of the four electromagnetic form factors of the trinucleon. (There are four since the trinucleon has spin and isospin each $\frac{1}{2}$). I emphasize the electric rather than the magnetic form factors, to minimize non-additive effects. The slope of an electric form factor at $q^2 = 0$, when plotted against the squared momentum transfer q^2, gives us the mean square radius of the charge distribution. This is calculated in a straight-forward manner from the trinucleon wave functions found by the different methods discussed above. Table 6.4 gives various calculations of the mean square charge radius for the triton and ^{3}He, along with the

Table 6.4. Trinucleon properties, for Reid soft core

Method of calculation	$r(^3\mathrm{H})$	$r(^3\mathrm{He})$	$P_{S'}$	P_D	Reference
Unitary pole approx.	1.76	1.97	—	—	Hadjimichael et al. (1972)
2-dim. Faddeev, mom. space	1.80	2.05	1.6	8.5	Malfliet (1972)
2-dim. Faddeev, mom. space	1.86	1.96	1.7	8.6	Harper et al. (1972)
Faddeev, co-ord space	1.65	1.90	1.8	9.0	Laverne (1973)
Variational	1.78	—	1.	9.5	Hennell (1973 A)
Hyperspherical harm.	—	—	0.4	8.9	Demin et al. (1973)
Experiment	1.70 ± 0.05	1.88 ± 0.05	—	—	Collard et al. (1965)

The charge root mean square radii in fm include the effects of nucleon electric form factors, $P_{S'}$ and P_D are the probabilities, in percent, of the S' and D states of the trinucleon, respectively.

experimental values. In making this comparison, we include the effects of the mean square radii of the nucleons, assuming that the charge distribution inside each nucleon is independent of the nucleon position: i.e., that we neglect non-additive effects discussed in Chapter I as part of the Gordian knot. The table shows fairly good agreement among the different calculations, and between calculated and experimental values.

We also compare the percentages of mixed symmetry S' state, and D state, for the calculated wave functions. Here, unfortunately, the experimental values are not well determined. The S' and D states affect the magnetic moments, but so do the not too well known exchange current (or isobar) effects. The agreement among the calculated values is good for the D state, but the S' probabilities vary by a factor of four.

The calculated differences between the root mean square radii of ^{3}He and ^{3}H are due to two effects acting in the same direction. First, ^{3}He contains an extra proton, which is much larger than the extra neutron in the triton. Second, there is an interference term between the completely symmetric S state (for spatial exchange of any two nucleons) and the S' state of mixed symmetry. For symmetric and mixed symmetry states with the same phase, this interference term also makes ^{3}He larger than the triton. The calculations usually do not take account of the mass splitting in the isospin doublet (due to coulomb forces and the charge asymmetry of nuclear forces as discussed above) which also would make ^{3}He the larger of the doublet. Finally, all the calculated values use trinucleon wave functions which give too high a value for E_T (Table 6.2) and therefore tend to give too large a value of the mean square radius for each member of the isodoublet. Apparently these last two effects are not very large, since (unlike the deuteron) the root mean square radii of the trinucleon are not greatly influenced by the asymptotic wave-function at large separation of a single nucleon from the center of mass of the other pair which is determined by the separation energy for that nucleon.

Schiff (1964) calculated the electric form factors neglecting non-additive effects. The triton form factor F_H is

$$F_H = 2 G_{En} F_L + G_{Ep} F_0 \tag{6.25}$$

while the ^{3}He form factor F_{He} is found by interchanging neutron (n) and proton (p) for the nucleon electric form factors G_{En} and G_{Ep}

$$F_{He} = 2 G_{Ep} F_L + G_{En} F_0 . \tag{6.26}$$

The form factors based on the trinucleon wavefunction are F_L for the *like* nucleons, and F_0 for the odd nucleon

$$F_L = F_1 - F_2/3 ,$$
$$F_0 = F_1 + 2 F_2/3 . \tag{6.27}$$

Here F_1 uses the spatially symmetric wavefunction u for the S-state

$$F_1(q) = \int \exp(iq \cdot r_1)\, u^2\, d^3 r_i . \tag{6.28}$$

The interference term uses the mixed symmetry functions v_1 and v_2 for the S' state.

$$\begin{aligned} F_2(q) = \int &\{[\exp(iq \cdot r_1) - \exp(iq \cdot r_3)]\, u v_1 \\ &+ 3^{\frac{1}{2}} \exp(iq \cdot r_2)\, u v_2\}\, d^3 r_i . \end{aligned} \tag{6.29}$$

Several workers have used these equations, and their expressions for the trinucleon wavefunctions u, v_1 and v_2 for the Reid potential to evaluate the form factors F_H and F_{He} for comparison with experiment (Collard, 1965; McCarthy, 1970; Bernheim, 1972). The scattering data (Collard, 1965) for the triton extends only to squared momentum transfer q^2 of 8 fm^{-2}. The form factors for both nuclei for $q^2 \leqq 8$ fm^{-2} are in general reproduced well by the calculated values. Measurements on ^{3}He up to q^2 of 20 fm^{-2} show a dip (not necessarily a zero) near 12 fm^{-2}, and a flat secondary maximum at higher values. Figure 6.7 compares measurements of F_{He} with the calculations of Gignoux et al.

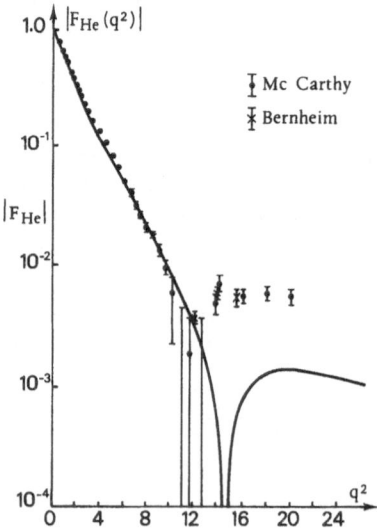

Fig. 6.7. Semi-logarithmic plot of the absolute value of the electric form factor of ^{3}He vs. squared momentum transfer in fm^{-2}. The dots show measurements of Collard et al. (1965) or McCarthy et al. (1970); the crosses show measurements by Bernheim et al. (1972). The solid curve shows calculations by Gignoux et al. (1972) for the Reid soft core potential.

(1972). Other calculated curves (Tjon, 1970, Yang, 1971; Hadjimichael et al., 1972) are very similar to Gignoux's up to q^2 of $10\,\mathrm{fm}^{-2}$, and differ somewhat in the position of the calculated diffraction minimum and of the height of the secondary maximum. The good agreement between all calculations and experiment for low values of q^2 is to be expected since all curves *must* agree exactly at $q^2 = 0$, and also agree in the charge mean square radius, Table 6.4. There is serious disagreement in the region of q^2 from 8 to $20\,\mathrm{fm}^{-2}$, and there has been considerable discussion both concerning the position of the diffraction minimum and the height of the second maximum.

Neither the experimental nor the theoretical situation is clear for the $^3\mathrm{He}$ electric form factor F_{He} at high momentum transfer. The main experimental problem is that the quantity measured is $(F_{\mathrm{He}})^2$. At low momentum transfer F_{He} is certainly positive, since we know its static value from the electric charge. With sufficiently accurate data, one can hope to follow $(F_{\mathrm{He}})^2$ to the diffraction minimum where it goes through zero: if in fact it does. Here insufficient attention is paid to the recent measurements of Bernheim et al. These data are more accurate than the earlier work, and are concentrated in the region of the "diffraction minimum". Even with this more accurate data, I do not see an unambiguous interpretation. It is clear that the diffraction minimum is *not* at the frequently quoted $11.8\,\mathrm{fm}^{-2}$, since a Bernheim measurement shown at that value is significantly different from zero. The combined data *suggest* a diffraction minimum at $11.0\,\mathrm{fm}^{-2}$; but given the ambiguity of sign of F_{He} and the experimental error other interpretations are possible. In any case, there is a clear disagreement between the calculated curve and the data for q^2 greater than $16\,\mathrm{fm}^{-2}$, where the calculated first maximum in the diffraction pattern is a factor of at least three below the experimental form factors.

Brayshaw (1973A) and Ballot et al. (1972) have independent arguments showing that the form factor measurements, and the triton energy, cannot be fitted with a pure two-body force compatible with measurements on nucleon-nucleon scattering. Brayshaw uses the measured form factor to find the function $\psi_0(r)$ for the lowest hyperspherical harmonic. He substitutes this function into (6.17), assuming a single uncoupled equation ($V_4 = V_6 = 0$) and determines the lowest multipole $V_0(r)$ as a function of the hyper-radius r. He finds that $V_0(r)$ agrees with that calculated for the usual two-body forces (averaging spin singlet and spin triplet) using (6.19), for hyper-radius r greater than 1.4 fm. But his $V_0(r)$ is more attractive than the standard values at smaller values of the hyper-radius. Brayshaw interprets this discrepancy as evidence for short range attractive three-body forces, which would contribute to $V_0(r)$ determined from the experimental form factor.

Ballot et al. (1972) solve the trinucleon problem for a variety of spin dependent central two-body forces using coupled hyperspherical harmonics. The potentials are adjusted to fit both the triton energy E_T, and the electric form factors from electron-trinucleon scattering. They are able to make such a fit, only if they ignore two-body scattering data. Their two-body force, designated "$G2$" has a strong and large repulsive core: so strong that the nucleon-nucleon ^1S phase shift goes through zero at the low energy of 160 MeV, instead of the experimental 260 MeV.

Both papers show that *some* new effect must be considered besides non-relativistic two-body forces, adjusted to fit nucleon-nucleon scattering. Brayshaw concludes that three-body forces are important at small values of the hyper-radius. I feel that it is hard to draw firm conclusions at this time, since all the difficulties listed above concerning E_T are likely to be serious, and we have the extra difficulty of non-additive effects in the electric form factors. i) The accuracy of the calculations at very small distances or high momenta (of order $3\,\text{fm}^{-1}$) is subject to serious doubt. ii) The two-nucleon potentials are poorly known at these small distances (of order $\frac{1}{3}\,\text{fm}$) and the extrapolation of the t-matrix to far-off-shell values needed in the Faddeev equation introduces great uncertainties. iii) Relativistic effects are particularly important at large values of the momentum. iv) The poorly known three-body forces are believed to be of quite short range (see Fig. 6.6), so are likely to have large effects on the part of the trinucleon wave-function giving us the form factor at large q^2. v) Exchange or isobar contributions to the electric form factor are likely to be of importance at high q^2-particularly in the region of a diffraction minimum.

It is certain that we have a lot to learn from the trinucleon form factors. But it seems hard to give a firm interpretation of the trinucleon form factors, before a similar analysis is successful in the (theoretically) simpler problem of the deuteron form factor discussed in Chapter 2. The accident that the deuteron has spin one and the trinucleon spin $\frac{1}{2}$ has made the experimental determination of the trinucleon form factors very much simpler than for the deuteron where the $G_0(q)$ and $G_2(q)$ are scrambled. So we do not yet have experimental curves of the monopole deuteron form factor $G_0(q^2)$ with a position of the diffraction minimum and height of the second maximum, to be explained by the use of a specific nucleon-nucleon potential, and a specific model for non-additive effects. After these experiments and their analysis is successful, we can return with some hope that we can make an unambiguous analysis of the trinucleon form factors at large q^2. We will still have three new features in the trinucleon problem: different two-body off-shell effects; the much discussed three-body forces; and the different non-additive effects for the trinucleon. But one can hope to disentangle these three effects.

6.11 Summary

There are now several excellent methods of calculation of the trinucleon ground state for a central, spin independent, local potential with a soft core. There are disagreements of about an MeV for calculations on the trinucleon energy for the Reid soft core non-central potential, but most calculations are near -7.1 MeV. The calculated values of E_T are about 1.4 MeV higher than experiment. This discrepancy is likely due to the choice of potential, to relativistic effects, and to three-body forces. Present calculations give reasonable agreement with each other and with experiment on the trinucleon root-mean square charge radii and form factors for electron scattering, up to q^2 of about 8 fm^{-2}. Disagreements at higher q^2 are likely due to the three effects listed above, and also to non-additive contributions to the electric form factors.

VII. Nucleon-Deuteron Scattering, and Other Problems

7.1 Introduction

In the last two chapters, I presented a variety of methods of calculation of the energy E_T of the trinucleon ground state. I also compared the calculated value of E_T and of the electric form factors for electron-trinucleon scattering with experimental data. In this chapter I present a summary of calculations involving continuum states of the three-nucleon system: I concentrate on the problems on nucleon-deuteron scattering (elastic and inelastic). I then turn to calculations on heavier nuclei: the alpha particle, the five-body system, etc. In principle I could discuss excited discrete or continuum states of A nucleons (with A at least 4), but there has been little progress to date based on attempts at reasonably exact solutions of the Schrödinger equation, so I omit this very large research area. (That is, I do not discuss the "less fundamental" approaches for scattering or reactions, such as quasi-free scattering, stripping, pick-up, Glauber approximation, etc.) I then return to the problems of unravelling, or cutting, the Gordian knot, discussed in Chapter I and throughout this work.

One purpose of my broad discussion, with few details, is to present a guide to the current literature in this area in which there is a wealth of experimental data, and rapid progress in theoretical work. A second more subjective purpose is to attempt to evaluate the different calculational methods, so that one might guess which methods would likely prove more successful in which type of problem. Obviously these guesses are not reliable. They amount to my "thinking out loud" as to the direction I expect that *my* research will take in the next several years.

I am not bold enough to try to plan the research programs for the entire community of nuclear theorists who are working on problems of few-nucleon systems.

Before starting continuum states of the three-nucleon system, I note that there is general agreement that there are no particle-stable excited states: i.e., no states in the energy range $E_T < E < -B$. Here, as in Chapter V and Chapter VI, energies are measured in the center of mass system, with zero corresponding to three free nucleons with negligible kinetic energy. No such state has been found experimentally. The theoretical situation is of interest, because there *would* be such a particle-stable excited state, for a central *spin-independent* nucleon-nucleon potential. For instance, Humbertson et al. (1968) found such states for a local potential, provided that the nucleon force was strong enough to bind the deuteron. But when the force is spin-dependent, the energy difference between the ground state and the first excited state increases, and for sufficient spin-dependence the higher state remains in the continuum (Badalyan et al., 1969).

Several years ago Efimov (1970) proposed a curious effect in a three body system composed of three identical bosons (or fermions with spin-independent forces). Namely, when the two-body attractive force is just large enough to give a two-body resonant state at zero energy (or an infinite scattering length), then the three-body system has in addition to its ground state a logarithmically infinite number of excited states with very small negative energy. This series limit at zero energy is reminiscent of the well known behavior of the energy levels of the hydrogen atoms, and in fact both series limits are due to a long range force. Amado (1972) reviews several of his calculations giving a quantitative demonstration and analysis of the Efimov effect.

The Efimov effect is of considerable theoretical interest. But it does not affect the energy levels for the triton, due to the strong spin-dependence of the nucleon-nucleon potential (Efimov, 1972).

7.2 Survey of $n-d$ Scattering

I discuss nucleon-deuteron scattering in terms of the schematic diagram Table 7.1. The energy of the system E_N increases upward. For comparison with the usual notation for scattering, I make two changes in our previous notation for energy. First, our new zero is defined for a nucleon-deuteron system with negligible kinetic energy. Second, the nucleon energy is now given in the laboratory system; non-relativistically the nucleon energy in the laboratory system is $\frac{3}{2}$ that of the energy in the center of mass system. One significant point on our energy scale is a

Table 7.1. A schematic for nucleon–deuteron scattering

Energy	Elastic			Polarizations			Break-up		
	σ_{tot}	σ_e	$d\sigma/d\Omega$	nucleon	d(vector)	d(tensor)	σ_B	triply diff.	Polarizations
very high									
pion thres.									
high									
moderate									Fig. 7.6	
low			Fig. 7.4 Sec. 7.4				Fig. 7.5 Sec. 7.5		
break-up threshold							0	0	0
very low	Fig. 7.2 Sec. 7.4 (^2a and ^4a)		Fig. 7.3 Sec. 7.3			0	0	0	0
0			Fig. 7.1	0	0	0	0	0	0

σ_{tot}, σ_e and σ_B are the total, elastic, and break-up cross sections, for nucleon–deuteron scattering; $d\sigma/d\Omega$ is the differential cross section for elastic scattering; the recoil deuteron (d) has vector and tensor polarization.

nucleon laboratory energy E_B of 3.33 MeV, corresponding to the center of mass energy of 2.22 MeV, allowing for deuteron break-up. Below this threshold, nucleon-deuteron scattering is completely elastic (neglecting the extremely small cross section for bremsstrahlung). At higher energies we have both elastic scattering and break-up. The latter produces a final state in which, asymptotically, we have three free nucleons. At still higher energies we reach the threshold for pion production. In the non-relativistic approximation this is at 210 MeV; in practice the effective threshold is higher since production of low energy pions goes principally through the intermediate state of the Δ isobar, some 300 MeV (center of mass system) above the nucleon mass. We limit our discussion to energies below the threshold for pion production since we wish to avoid final states of *four* hadrons. Also in practice the calculations become more difficult as the energy increases; so with present techniques we stop at some modest energy of order 100 MeV.

The quantities measured become more and more complicated as we move to the right in Table 7.1. The total cross section σ_t is a single number at a given nucleon energy, which can be measured by a "sample in-sample out" measurement to an accuracy of order 1 %. (For protons we must, of course, exclude the coulomb scattering in the $p - d$ system, which gives an infinite value for σ_t. Though I generally refer to "nucleons" I am always thinking of "neutrons" for calculated nucleon-deuteron scattering, both for this reason, and because most of the calculations do not treat protons. This leads to a paradox, since many of the detailed measurements in Table 7.1 involve $p - d$ rather than $n - d$ scattering because it is much easier to get collimated, monoenergetic and polarized beams of protons). When we go to nucleon energies higher than the break-up threshold E_B, the total cross section splits into two parts: the cross section for elastic scattering σ_e, and the cross section for break-up σ_B (σ_B in turn splits into two parts above the threshold for pion production).

The next several columns, going to the right in 7.1, show more details in elastic scattering. First, we have the column for the differential cross section $d\sigma/d\Omega$, for elastic scattering, with no measurements of nucleon or deuteron polarization. Now a single block in our table represents a whole curve of the differential cross section vs. scattering angle θ, at a specified nucleon energy.

Continuing to the right with columns for elastic scattering, we come to measurements on polarization effects in nucleon-deuteron scattering. An incomplete, but usual, set of measurements consists of measuring three different polarization parameters for an initially unpolarized system: the nucleon polarization P_N, the deuteron vector polarization P_V^D, and the deuteron tensor polarization P_T^D (e.g. McKee et al., 1972).

Each of these polarizations is a function of the scattering angle θ, and the nucleon energy E_N.

Going still further to the right in Table 7.1, we come to the column for the integrated break-up cross section σ_B. We then come to highly multiple differential cross sections, such as $d^3\sigma/dE_1 \, dE_2 \, d\Omega_1$ or $d^3\sigma/dE_1 \, d\Omega_1 \, d\Omega_2$ where the subscripts designate the nucleon whose energy E or angle θ is measured. Going still further, we come to polarization measurements in elastic scattering, and even to the analogue of triple-scattering parameters in nucleon-nucleon scattering. [See Hoffman et al. (1962) for polarization measurements at 140 MeV. There are few such measurements at lower energy.]

When we compare this enormous wealth of data with the single experimental number E_T for the triton energy – which took us two chapters to calculate – we get a feeling for the richness of the research area opened up when we turn to continuum states. It is, in fact, even richer, since we can reach these continuum states by a variety of other means: e.g., by the trinucleon photo-effect leading to either nucleon-deuteron or three nucleon systems (Barbour et al., 1970), or by inelastic electron-scattering, in which we measure inelastic form factors (Lehmann, 1969).

Experiments have been made for most of the blocks in Table 7.1. I give references to figures, or sections of this chapter in the different blocks.

Calculations are still in a rudimentary stage for most of the table: the lower left hand block for elastic scattering below the break-up energy E_0 is in fairly good condition, and preliminary attempts have been made throughout the table. The absence of reliable calculations provides a brake on the experimental work, especially at the far right where the wealth of data makes it very hard to present results in a digestible form, even if one succeeds in making the measurements.

7.3 Neutron-Deuteron Scattering Lengths

In contrast to the experimental determination of the trinucleon energy E_T, the experiments involving continuum states are far from trivial. I start with a review of the history of experiments for the bottom row of Table 7.1: i.e., for negligible nucleon energy. In this case, the scattering must be elastic, and should involve only S-wave neutrons. (We shouldn't have a resonance for a state of higher angular momentum right near zero.) We must use neutrons to avoid large coulomb effects. Since the neutron and deuteron have spins of $\frac{1}{2}$ and unity, the total spin S (in Russell-Saunders coupling notation) can be either $S = \frac{1}{2}$ or $S = \frac{3}{2}$. The scattering lengths for these two spin states are written ^2a, and ^4a, respectively; and the entire row reduces to the determination of these two scattering lengths.

Delves et al. (1969) review the earlier history, which involved the resolution by Alkimov et al. (1967) of the ambiguity of two quite different sets of solutions usually designated as A and B. The set B was eliminated, and subsequent work involved more precise determinations of the scattering lengths for set A. Delves gives

$$^2a = 0.15 \pm 0.05 \text{ fm}; \quad ^4a = 6.13 \pm 0.04 \text{ fm} . \tag{7.1}$$

Of course, two independent measurements are needed to give the results, 7.1: namely the amplitude for coherent $n - d$ scattering (Bartolini et al., 1968) and the spin-incoherent cross section (Gissler, 1963). Equation 7.1 gives $\sigma_e(0) = 3.15 \pm 0.04$ b.

In the past several years, new measurements have been made both for $\sigma_e(0)$ and for coherent scattering. Dilg et al. (1971) measured the total cross section for 130 eV neutrons and Stoler et al. (1972, 1973) measured it at a continuum of neutron energies, from 1 keV to 200 keV. Their results $\sigma_e(0) = (3.39 \pm 0.01)$ b are consistent with each other, and with the early Fermi (1949) measurements. Dilg also measured the amplitude for coherent scattering. Combining these two measurements, he finds the scattering lengths

$$^2a = 0.65 \pm 0.04 \text{ fm}; \quad ^4a = 6.35 \pm 0.02 \text{ fm} . \tag{7.2}$$

There is still some controversy; but I guess that the results (7.2) are likely to survive, in preference to 7.1.

Calculations of the scattering lengths have been made for a number of years by various methods for several different potentials. See Duck (1968); Delves et al. (1969); Mitra (1969); Amado (1969); and Sitenko et al. (1971). First consider a spin-independent separable approximation. The Faddeev equation (5.22) for the bound state is modified to include an inhomogeneous term for calculation of the scattering lengths. The calculations for a tensor separable potential is as simple for calculations of the scattering length as for the trinucleon energy, and a large number of such calculations have been made for different separable potentials. Delves et al. (1969) point out two simple ways of summarizing these calculations: 1, the quartet scattering length is almost independent of the assumptions made concerning the separable potential; 2. the doublet scattering length is sensitive to these assumptions, but there is a linear relation between the values of ^2a and of E_T for a variety of two-nucleon potentials that fit the low energy properties of the nucleon-nucleon system. Brady et al. (1970) have considered many more cases that also support this linear relationship, called the Phillips line (Phillips, 1968).

Barton et al. (1969) use dispersion theory to give a model-independent result for the quartet scattering length. The basic argument is that there is a long range repulsive force for this state. This long range can be understood from the relatively large size of the deuteron. The repulsive character is due to the Pauli principle, which prohibits having all three nucleons in a system with all spins pointing in the same direction, for two-body states with even orbital angular momentum for nucleon pairs. Barton uses dispersion theory to solve nucleon-deuteron scattering by the N/D method, for spin-dependent central two-body forces. He finds that the quartet scattering length is

$$^4a = -(\tfrac{3}{4}BM)^{\tfrac{1}{2}} N_4(0) = 6.3 \text{ fm}. \tag{7.3}$$

Here B is the deuteron binding energy, and M the nucleon mass. The numerical value $N_4(0) = -1.679$ is the result of his N/D calculation, and gives the value 6.3 fm quoted. He constrasts this $N_4(0)$ with the Born approximation value three times as large, $N_4^B(0) = -5.08$.

The value given by (7.3) is in remarkably good agreement both with the experimental (7.2), and with a variety of other calculations using the methods of Chapters V and VI. For instance, Sitenko et al. (1971) give values of ^4a of 6.336 fm for a Hulthén potential, 6.279 fm for a square well potential, and 6.35 fm for a Morse potential.

On the other hand, the value of ^2a is sensitive to the details of the nucleon-nucleon potential; but it seems to be sensitive to these details in just the same way as is the trinucleon energy E_T. The Phillips line relating ^2a and E_T for a variety of different choices of the potential is shown in Fig. 7.1. Harms (1972) gives the equation of the line as

$$^2a = 0.75\,(E_T + 8.5) + 0.75 \text{ fm}. \tag{7.4}$$

This way of writing the linear relationship shows that if the calculated trinucleon energy is near the experimental value, then the doublet scattering length of about 0.75 fm will also be near the experimental value (7.2). [For several years the existence of the Phillips line was regarded as a difficulty; since it showed that calculations could not agree with both the experimental energy, and the experimental value (7.1) for the doublet scattering length. But I adopt Eq. (7.2); then we automatically get a "good" value for ^2a.]

Why do calculations fall on or close to the Phillips line? I don't know. It is not clear to me whether *all* calculations for two-body potentials must do this (provided they agree with the low energy properties of the singlet and triplet nucleon-nucleon systems); or whether the closeness to the Phillips line is due to our selectivity in choice of nucleon-nucleon potentials. One can choose three-body potentials in a manner to stay on

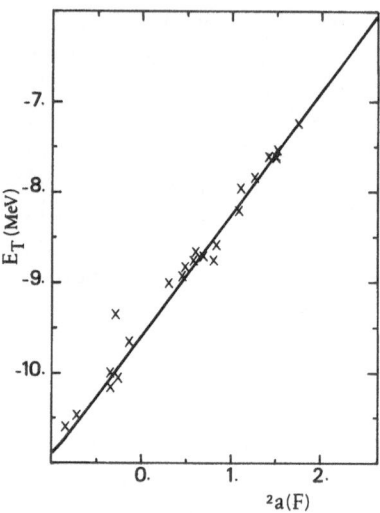

Fig. 7.1. Phillip's line of trinucleon energy E_T in MeV vs. doublet scattering length ^2a in fm. The points show results of calculations with a variety of separable potentials. The line obeys Eq. (7.4). From Brady et al. (1970)

the line; to go below it; or to go above it (Laroze, 1970). Thus the fact that experimental values lie very close to the line cannot be used to prove or disprove the importance of three-body forces. I have not seen a calculation concerning the possible shift from the line due to relativistic effects.

Since calculations for the Reid potential miss the experimental value of the triton energy, they also miss that for the doublet scattering length. *Presumably* the effects considered in Chapter VI that account for the disagreement between experimental and calculated values of the triton energy would, if we include them in a scattering calculation, give agreement with the experimental value (7.2).

The calculations of the scattering lengths are generally done with a separable approximation to the nucleon-nucleon t-matrix. However, some of the other methods discussed in Chapter VI have been used for scattering length calculations for local tensor forces: Malfliet and Tjon (1970) solve the two dimensional coupled Faddeev equations by iteration. Delves and Hennell (1971) find the scattering length for the Hamada-Johnston potential by a variational method; but I have not seen any results by this treatment for the Reid potential. I have not seen scattering length calculations by the Gignoux-Laverne (1972) solution of the Faddeev equations in coordinate space, by the use of hyperspherical harmonics (Simonov, 1972), or by the Bethe-Brueckner approach (Tripathi, 1972; Sotona, 1973).

7.4 Elastic $n-d$ Scattering

We have exhausted the bottom row in Table 7.1, so we continue to energies up to the break-up threshold, E_B. Again, calculations say for separable interactions can be made by the same general methods discussed above for the bound state, and for zero energy scattering. I shall first survey calculations made using a separable approximation to the nucleon-nucleon t-matrix. I then return briefly to calculations for local potentials, made by other methods.

In nucleon-deuteron elastic scattering we have the new feature that we must find the scattering amplitude for *each* partial wave of the two-body nucleon-deuteron system. Due to the large size of the deuteron, we should not be surprised at two effects which are qualitatively different from nucleon-nucleon scattering: 1. the S-wave scattering varies rapidly with neutron energy; 2. p-wave and higher partial waves are important at surprisingly low energies. Stoler et al. (1972) give calculations of these two effects for several different spin-dependent central separable potentials. Figure 7.2 compares their calculations with their measurements.

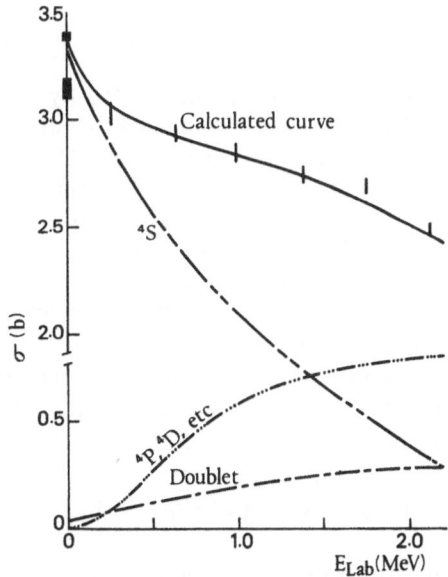

Fig. 7.2a. Total cross section for neutron-deuteron scattering, vs. neutron energy in MeV in the laboratory system. The four curves show calculations for spin-dependent separable potentials, for doublet states, for ^4P, ^4D, etc., ... states, for the ^4S state, and (solid) the sum. The vertical bars show experimental data (Seagrave, 1970) with errors. The small square at zero energy shows Fermi's and Dilg's measurements, and the solid rectangle shows Seagrave's extrapolated value. From Stoler et al. (1972)

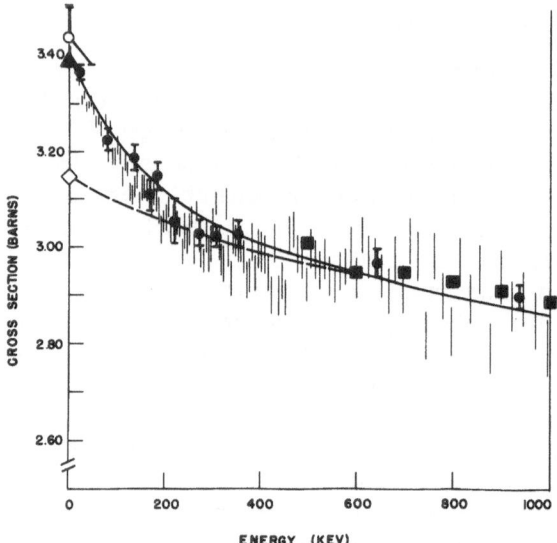

Fig. 7.2b. Total *n–d* cross section on an expanded scale. The solid circles, squares and vertical lines show experimental data; the solid calculated curve is redrawn from Fig. 7.2a. From Stoler (1973)

We see that the total cross section is dominated by the rapid decrease with energy of the scattering in the ^4S state, and the rapid rise with energy of the quartet scattering in higher partial waves. These two dominant effects happen to combine to give a cross section that varies slowly with energy in the range 0.4 MeV to 2 MeV, suggesting the smooth extrapolation that was made earlier to a zero energy cross section. In retrospect it seems surprising that the region below $\frac{1}{2}$ MeV was not explored both theoretically and experimentally, during the years when more difficult calculations were being made for higher energies.

For a central spin-independent separable potential, the Faddeev equations for elastic scattering can be written for the l'th partial wave in the form (Amado, 1969)

$$T_l(p, k; E) = B_l(p, k; E) + \int_0^\infty dq\, K_l(p, q; E)\, T_l(q, k; E). \tag{7.5}$$

Here $B_l(p, k; E)$ is the Born approximation to the scattering amplitude, and $T_l(p, k; E)$ is the scattering amplitude for the given partial wave, with momenta p and k, and energy E. The kernel $K_l(p, q; E)$ involves the two-body t-matrix and the centrifugal barrier. For energy E above the break-up threshold, the kernel has a branch cut, which makes this integral equation more difficult to solve than at lower energies. Most calculations to date for elastic scattering use spin-dependent separable potentials of Yamaguchi shape. The Faddeev equations then reduce to two coupled one-dimensional integral equations generalizing (7.5). Amado (1969) avoids the branch cut by use of the Hetherington-Schick (1965) method of contour-deformation. Instead of integrating along the real axis, he uses a contour in the complex q-plane that is chosen to avoid singularities. This method demands knowledge of the kernel, and therefore of the potential, for complex values of momenta q. Amado finds T_l for many values of l. I present two of Amado's early results (Aaron et al., 1965) for the differential cross section $d\sigma/d\Omega$ for elastic neutron-deuteron scattering: Fig. 7.3 and 7.4 for the differential cross section curves at neutron laboratory energies of 2.45, and 14 MeV respectively. The dashed line in Fig. 7.3 shows the gross failure of the Born approximation (cf. its failure for the quartet scattering length mentioned above); while the calculations shown as a solid curve are in good agreement with experiment (Seagrave et al., 1957) at large angles, but are about 20% low at small angles. Note that at 2.45 MeV, the scattering is highly anisotropic. Calculations are also in fair agreement with experiment (Allard et al., 1953; Seagrave, 1954; Berick et al., 1968) at a neutron energy of 14 MeV, illustrated in Fig. 7.4.

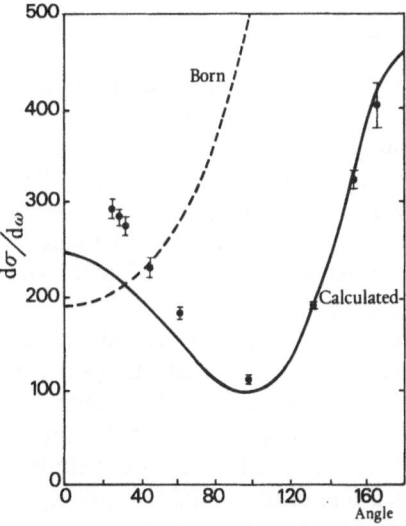

Fig. 7.3. Differential cross section for elastic neutron-deuteron scattering in mb/sr vs. neutron scattering angle in degrees in the center of mass system, at a neutron laboratory energy of 2.45 MeV. The solid curve is calculated from a spin-dependent separable potential; the dashed curve is the Born approximation; the points show experimental values. From Amado (1969)

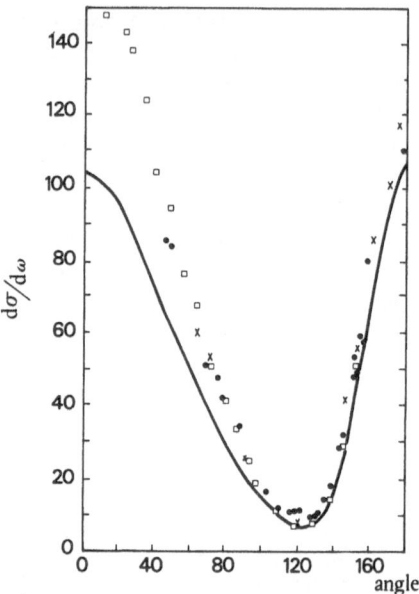

Fig. 7.4. Differential cross section for elastic neutron-deuteron scattering at a neutron laboratory energy near 14 MeV. The curve shows Amado's calculations; the points show several sets of experimental values. From Amado (1969)

The calculated imaginary part of the amplitude for elastic scattering at $0°$ is substituted in the optical theorem to give the total cross section σ_t. But the differential cross section for elastic scattering also gives us by integration the elastic scattering cross section σ_e. The difference $\sigma_t - \sigma_e$ gives a calculated value of the break-up cross section, σ_B. Figure 7.5 from Amado compares calculated and experimental values (Howerton, 1961; Catron et al., 1961) of σ_t and σ_B. Calculations and experiments agree to about 10% accuracy.

Much of the recent calculational work on elastic scattering consists of extensions of Amado's pioneering efforts. For instance, Laroze et al. (1970) calculated scattering in the $\frac{1}{2}^+$ state, using tensor forces. Aarons et al. (1972) and Doleschall (1972) treated other values of angular momentum and parity. Doleschall (1973) also includes two-body P waves, and solves the coupled one-dimensional integral equations by a different numerical procedure than the contour integration technique used by Amado and the others. Also calculations have been extended beyond 14 MeV. The results for the differential cross section for elastic scattering are similar to those we have just discussed. But with tensor forces, we can now hope to calculate polarization effects in nucleon-deuteron

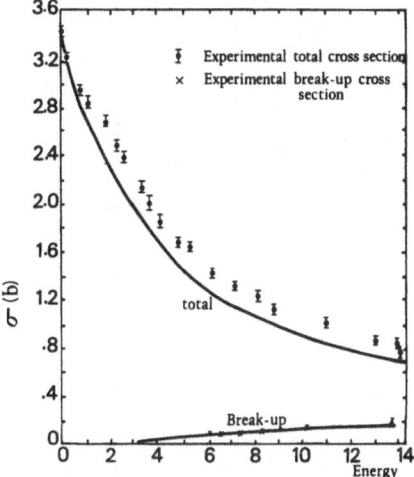

Fig. 7.5. Break-up and total cross sections for neutron-deuteron scattering, vs. neutron energy in MeV in the laboratory system. The curve shows Amado's calculations; the x's and circles show experimental data. From Amado (1969)

scattering: i.e., the nucleon polarization and the deuteron vector and tensor polarization in the schematic Table 7.1.

The situation is unsettled at present. There are three independent calculations of polarization effects in neutron-deuteron scattering for tensor forces with Yamaguchi form factors: by Aarons and Sloan (1972), by Doleschall (1972, 1973), and by Avishai-Rinat (1971). The first two agree with each other (Doleschall et al., 1972) but the third gives a different result: the reasons for this discrepancy are not known. The Sloan-Doleschall results are in only fair agreement with experimental measurements of the neutron polarization, and the deuteron vector and tensor polarization, even when two-body P-waves are included (Doleschall, 1973; Fiore et al., 1973). Pieper (1972 B, 1973) treats P-wave effects by perturbation theory, and finds agreement with experiment but *not* with Doleschall's calculations. Apparently perturbation theory is not adequate for this purpose. Further, the separable approximation used is known to give a poor approximation to the on-shell t-matrix for tensor forces: see our discussion in Section 4 concerning the Blatt-Biedenharn phase parameter δ_β.

7.5. Inelastic $n-d$ Scattering; New Methods

Calculations on inelastic nucleon-deuteron scattering (say of the triply differential cross section $d^3\sigma/dE_1\,d\Omega_1\,d\Omega_2$) have proved to be more

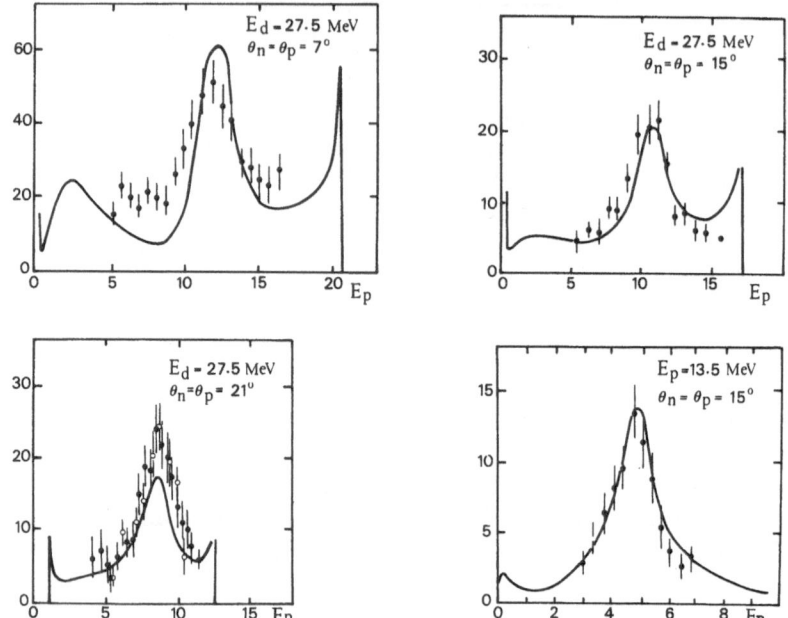

Fig. 7.6. Theoretical and experimental break-up spectra for $p-d$ scattering. The triply differential cross section in mb/sr^2 MeV plotted vs. proton energy in MeV for different energies and laboratory angles. The curves show Ebenhöh's calculations, compared with the experimental points. From Ebenhöh (1972)

difficult than those of elastic scattering, due to the singularities of the kernel. Calculations for spin-dependent central separable potentials of Yamaguchi shape made by Aaron et al. (1966) have been superseded by those of Cahill et al. (1971, 1972) and Ebenhöh (1972). The agreement with experiment (Burg et al., 1971) on proton-deuteron inelastic scattering shown in Fig. 7.6 is only fair. Ebenhöh points out that he has not included Coulomb effects in his calculation; but that it is uncertain whether these effects are the major source of the disagreement shown between his calculations and experiment.

I remarked above that rank-one separable potentials are likely to prove unsatisfactory for any continuum calculations involving polarization effects. We have two alternatives: first, to extend the calculations to treat rank-two separable potentials, or second, to develop new calculational methods that could apply to local potentials. Since the present calculations with rank-one separable tensor potentials are already difficult (they have not yet been done for inelastic scattering; and there are calculational disagreements for polarization effects for elastic scattering), an extension to higher rank is quite difficult. On the other hand, the second alternative of new calculational methods looks quite promising.

Several new methods that do not make use of a separable approximation to the t-matrix have recently been developed for calculation of nucleon-deuteron scattering. I shall name these methods; but I shall not present results, nor attempt any evaluation of their success. I have already referred to Pieper's use of Alt's (1967) perturbation theory. Pieper later (1973) used the Aarons-Sloan (1972) formulation of perturbation theory for scattering problems. A second approach is to apply variational methods; similar to those used for bound state problems. Delves et al. (1971) used these methods for the Hamada-Johnston potential, to find the scattering amplitude at nucleon energies below the threshold for break-up.

Pieper (1972A) and Brady-Sloan (1973) have applied a variational approach to nucleon-deuteron scattering. Harms (1972A, 1972B) has recently applied his "method of moments". Tjon (1970), Kloet-Tjon et al. (1971, 1973) use Padé approximants to find the amplitudes for elastic or inelastic scattering for the nucleon-deuteron system for a local central potential. Recently, Brady et al. (1972) checked the accuracy of this method, by comparing with known results for $n-d$ scattering for a separable t-matrix.

Very recently Brayshaw (1973B) has developed a new method for three-particle final states, using a boundary condition for the "interior region" (Noyes, 1969) together with fixed on-shell values of the two-body t-matrices. He applies (1973C) this method to find the triply differential break-up cross section at 14.4 MeV, and finds that the boundary condition constrains the cross section severely. A single adjustable parameter (the choice of boundary condition) fixes the triton energy, the doublet scattering length, and the triply differential cross section.

I turn briefly to the ^{3}He photoeffect, to illustrate the first use of hyperspherical harmonics for a continuum calculation. Fabre de la Ripelle and the present author (1973) treat electric dipole transitions from the ground state to excited 1^- states of isospin $\frac{3}{2}$ approximated as a single hyperspherical harmonic, with grand orbital of unity. The selection rule that the grand orbital changes by ± 1 selects out the two lowest values (0 and 2) of the grand orbital for the ground state wavefunction, (only $L=0$ is used for the published cross sections, Fabre and Levinger (1973)). The corresponding wave functions are taken from Ballot et al. (1972) for an assumed Volkov V^x potential. The continuum wavefunction is calculated as the solution of an uncoupled differential equation, using the nucleon-nucleon potential of Gogny et al. (1970). This simple calculation gives the remarkably good agreement shown in Fig. 7.7 with Fetisov's measurements (1965) on three-body break-up. [The isospin $\frac{1}{2}$ state makes

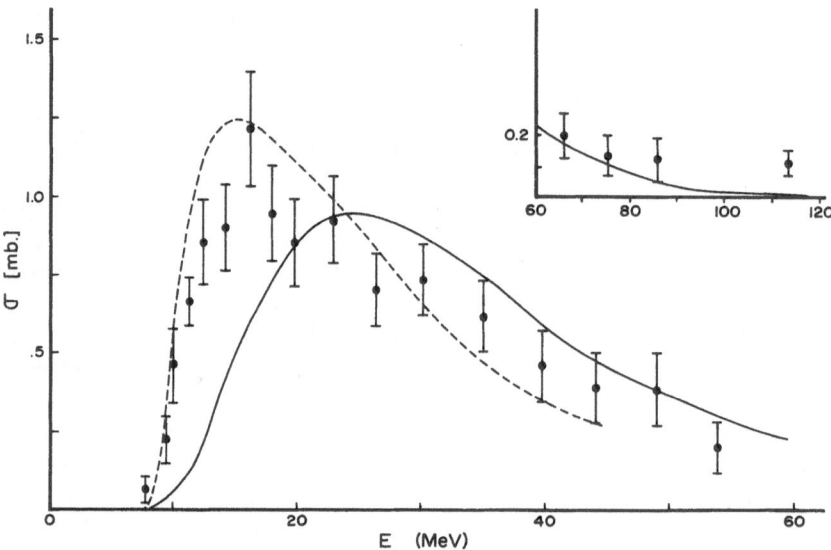

Fig. 7.7. Total cross section for ³He photoeffect. The crosses show Fetisov's (1965) experiments for three-body break-up; the dashed curve shows Barbour's (1970) calculations for a separable t-matrix; the solid curve shows Fabre's (1973) calculation for isospin $\frac{3}{2}$ using hyperspherical harmonics

only a small contribution to three-body break-up (O'Connell et al., 1969; Barbour-Phillips, 1970)], but *may* account for the serious disagreement with experiment from threshold to 18 MeV. I also show in Fig. 7.7 the calculation of Barbour and Phillips, based on a separable approximation to the two-body t-matrix. This approximation allows them to find the continuum wavefunction by solving one-dimensional integral equations. Barbour-Phillips treat both isospin final states ($\frac{1}{2}$ and $\frac{3}{2}$). It is necessary to check the supposed rapid convergence of h.h. assumed by Fabre-Levinger, by recalculating the photoeffect cross section using the final continuum state of isospin $\frac{3}{2}$ as a solution of N coupled differential equations. It remains unclear how to treat the isospin $\frac{1}{2}$ state by h.h., since here the nucleon-deuteron channel (which cannot couple to isospin $\frac{3}{2}$) is clearly of significance.

7.6 Alpha Particle, etc.

Continuing our survey of other problems in nuclear physics, I turn to the problem of calculations of the ground state energy for systems with more than three nucleons. (We could later include the rest of nuclear physics by treating scattering problems, for systems of more than three nucleons; but unfortunately we will not be able to get that far.) As in Chapter V, use of a separable approximation in the Faddeev equations lead to a reduction by unity of the dimensionality of the resulting integral equations; but in this case the remaining equations are still very difficult. Consider the simplest case of the alpha particle for central, spin-independent forces with a separable t-matrix. The trinucleon gave us the single one-dimensional integral equation (5.22) which is easy to solve numerically. But Panschapakesan (1965) found that the alpha

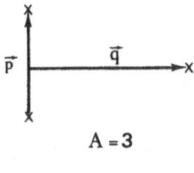

A = 3

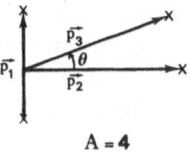

A = 4

Fig. 7.8. Vectors needed in the three-nucleon problem (above) and the four-nucleon problem (below)

particle gives us a single *three-dimensional* integral equation for this oversimplified case. A heuristic way of reaching this result is illustrated in Fig. 7.8. For the triton, we need the momentum coordinates p and q, and express the Faddeev equation in terms of the pair function $\psi(p, q)$ of Eq. (5.18), which reduces to a one-dimensional integral equation when we make a separable approximation. In the alpha particle, we have *three* vectors, p_1, p_2, and p_3. With a separable approximation we can factor out the dependence on p_1; but we are left with a dependence on the magnitudes p_2 and p_3, and on the angle θ between them: hence three variables.

There has not been a serious effort to solve this three-dimensional integral equation, or à fortiori the coupled three-dimensional integral equations resulting from spin-dependent and/or non-central forces. With recent advances in numerical techniques and in computers, it seems not impossible that one might solve the alpha particle using the approach of Chapter V. But the problem gets worse rapidly as A increases, so it is clear that direct solution of the Faddeev equations for separable potentials soon comes to an end with increasing mass number. This conclusion holds, a fortiori, for direct solutions of the still higher dimensional Faddeev equations for a local potential.

Which of the other methods I discussed in Chapter VI are applicable to the problem of the alpha particle, and heavier nuclei? While there have been many variational calculations for the alpha particle, I have not seen any similar to those of Delves et al. (1971) or Hennell et al. (1972) for the triton: i.e., with two-body non-central forces with strong short-range repulsion. Tripathi et al. (1972) have applied the Brueckner approximation to the alpha particle and finds an energy of -17.35 MeV for the Hamada-Johnston potential. As in his work on the triton, problems arise concerning the accuracy of the Brueckner approximation, and concerning the values of certain parameters assumed.

On the other hand, the use of the optimal subset of potential harmonics (Demin, 1972, 1973) provides just about as rapid convergence for the alpha particle, as that discussed in Chapter VI for the triton problem. I have not seen any calculations on the alpha using the Reid potential. But Demin et al. (1973) demonstrate the convergence of the h.h. expansion for the EH (Eikemeier, 1971) and GPT (Gogny, 1970) potentials. These potentials include tensor forces and repulsion at short range, but they do not have as strong repulsion as that used by Reid, in his "soft core" fit. Both potentials show good convergence at $k_m = 14$. The energy values E_α depend, of course, on the two-body potential assumed; as they also do for the triton results E_T. I include results discussed above for the Hamada potential.

$$E_T \text{ (Hamada)} \quad = - \ 6.5 \ \text{MeV (Hennell)}$$

$$E_T \text{ (Reid)} \quad = - \ 6.6 \ \text{MeV (Demin)}$$

$$E_T \text{ (Eikemeier)} \quad = - \ 7.45 \ \text{MeV (Demin)}$$

$$E_T \text{ (Gogny)} \quad = - \ 8.61 \ \text{MeV (Demin)} \qquad (7.6)$$

$$E_\alpha \text{ (Hamada)} \quad = - \ 17.35 \ \text{MeV (Tripathi)}$$

$$E_\alpha \text{ (Englekmeier)} = - \ 23.05 \ \text{MeV (Demin)}$$

$$E_\alpha \text{ (Gogny)} \quad = - \ 26.76 \ \text{MeV (Demin)}.$$

We see that the effect of the Hamada-Johnston hard core of 1 MeV on the triton energy seems to be increased by an order of magnitude for the alpha particle energy. (The quantitative comparison of Tripathi's and Demin's calculations cannot be taken seriously at present; but the order of the effect is likely correct.)

One is tempted to extrapolate and conclude that hyperspherical harmonics provide a good method of calculating ground state energies for still larger nuclei. In fact, only very primitive calculations have been made to date, beyond mass 4. Simonov (1972) points out that we do not yet have expressions for the optimal subset for larger nuclei; and that it is excessively difficult to include enough hyperspherical harmonics unless the optimal subset is used.

A somewhat different approach can be used in studying larger nuclei: they can be regarded as collections of a small number of smaller nuclei. For instance, Harrington (1966) has calculated the properties of ^{12}C, regarded as composed of three alpha particles. He used separable t-matrices, adjusted to agree with alpha-alpha scattering; but one could easily use the methods discussed in Chapter VI for a local alpha-alpha potential. Appreciable Coulomb forces provide some extra trouble. The most serious difficulty in this calculation is that, of course, the alpha is not really a "particle". (But then the nucleon isn't either; through we have to go to higher energy to find that out!) The internal structure of the alpha implies that we should expect to find appreciable "three-body forces" in the carbon nucleus. If we had a good method of calculating the distortion of one alpha particle by another, and the effect this has on the forces on a third alpha, then we could correct for these three-body forces.

A somewhat different question concerns the calculation of the properties of hypernuclei, such as $^{3}_{\Lambda}$H, composed of a neutron, a proton, and Λ hyperon (Pniewski, 1972). The methods using the Faddeev equations for separable interactions have been adopted to this problem, with different masses, and with forces depending on hypercharge. (Hetherington, 1965; Dabrowski et al., 1968; Hartt, 1971). Variational methods have been used by Downs (1966), Bodmer (1966) and Herndon-Tang (1967). One could also apply hyperspherical harmonics to a problem where the three particles have different masses, by modifying the definitions of the variable ξ_1 and ξ_2. I have not seen these h.h. calculations; they would be of particular interest for $^{4}_{\Lambda}$H and $^{4}_{\Lambda}$He, where the other calculational methods are less satisfactory. Of course, the Λ-nucleon force is known to a considerably lower accuracy than is the nucleon-nucleon force; the main motivation in studying hypernuclei is to learn more about the $\Lambda - N$ force. Present calculations neglect the possible $\Lambda - N$ tensor force: this approximation *may* not be too serious, as the long range part of the tensor force, provided by OPEP in the $N - N$ case, is now absent due to isospin, since the Λ has isospin zero. Recently Dabrowski (1973) has treated the coupling of the Λ to the Σ hyperon, only 60 MeV more massive. This admixture of higher states is similar to Δ-admixture in a nucleon-nucleon system.

7.7 Comparison of Different Methods of Calculation

I present in Table 7.2 a schematic survey of the present status of different calculational techniques to a variety of problems on few-body systems: the ground state of the trinucleon, nucleon-deuteron scattering, the ground state of the alpha particle, heavier nuclei, and hypernuclei such as $^3_\Lambda$H. I assume a local potential, such as Reid's soft core. My entries "good", "ok", "likely", maybe", and "unlikely" are clearly subjective: they are a function of the individual making the entry, and the time it is made. The work "good" means that an accurate calculations has been performed. (Any disagreement with experiment is then to be blamed on the Hamiltonian used – not on inaccuracies or approximations in the calculation. Since some columns, such as $n-d$ scattering, include many different calculations as illustrated in Table 7.1, the "good" calculations may well be incomplete.) The noncommittal word "ok" means calculations can be made; but the approximations may be inaccurate: e.g., the use of the UPA for the trinucleon ground state. If I believe a method would apply ok, but I know of no such calculations, I write "likely". If a formalism has been developed, but not tested yet, I use the noncommittal work, "maybe". If I think a calculational method would fail if tried, I write "unlikely". If a method hasn't been tried, and I have no basis for guessing whether it would work or not, I write "?".

The above discussion is all for a local potential. There has recently been considerable interest in the use of phase-equivalent non-local (and non-separable) potentials. The following calculational methods, as currently used for local potentials, are not applicable to non-local potentials: Faddeev equations in coordinate space, and hyperspherical harmonics. It seems likely that these formalisms could be developed to work with non-local potentials, also.

A glance at the table shows that there is no one "universal method" for solving the problems of nuclear physics. We have *many* good methods of solving the trinucleon ground states. Many methods have been applied to nucleon-deuteron scattering (from low energy to the threshold for pion production); but it is too soon to specify which methods are completely successful. So far only hyperspherical harmonics and the Brueckner-Bethe method have been successfully applied to the alpha particle. Very little has been accomplished to date on heavier nuclei; while light hypernuclei such as $^3_\Lambda$H have been calculated by several methods.

I divide future work into four main categories: 1. further calculations on the trinucleon ground state; 2. nucleon-deuteron scattering, with more and more complex quantities calculated and measured; and 3. the

Table 7.2. Applicability of different calculational techniques for the Reid potential

Method	Section	Trinucleon ground state	Nucleon-d scattering	Alpha	$A > 4$	$^3_\Lambda H$
Separable approx.	Ch. 5, 7.4	OK	OK	unlikely	unlikely	OK
Separable approx. + perturb. theory	6.2	good	OK	unlikely	unlikely	likely
Faddeev equations, Padé approximation	7.5	good	good	unlikely	unlikely	likely
Faddeev equations moments	7.5	—	maybe	unlikely	unlikely	—
Faddeev equations, coordinate space	6.3	good	?	unlikely	unlikely	unlikely
Variational	6.4, 7.5	good	OK	unlikely	unlikely	OK
Hyperspherical	6.5	good	?	OK	maybe	likely
Brueckner-Bethe	6.6	OK	?	OK	OK	maybe

For further discussion of the subjective judgements, "good", "OK", "likely", "maybe", "?", and "unlikely" see Section 7.7.

alpha particle; and 4. the rest of nuclear physics. I shall attempt to evaluate various calculational methods for these four categories.

Non-relativistic calculations for the trinucleon ground state with an *assumed* two-nucleon potential are now in quite good shape. Since a variety of reliable methods are now available; the choice as to which to use is principally one as to which programs one has for the computer at hand, and the speed and memory of the computer. The unitary pole approximation (or Afnan's "1 A" term for tensor forces) with corrections using perturbation theory, might continue to be useful here as it is a relatively fast and cheap calculation, allowing the study of many different potentials.

But the unitary pole approximation is less successful for scattering problems, by direct solution of the Faddeev equations in momentum space: e.g., it has not been applied to break-up calculations including tensor forces. Also quantities such as polarization in $n - d$ scattering are more sensitive than the trinucleon energy to t_{02} and t_{22}, where the unitary pole approximation is inaccurate. On the other hand, new methods using Padé approximants, variational calculation of scattering, or solution of the Faddeev equations by the method of moments, work about as well for local as separable potentials. While Badalyan et al. (1972) have developed formalisms for treating problems such as $n - d$ scattering using expansions in hyperspherical harmonics, calculations have not yet been carried out using these techniques. The isospin $\frac{3}{2}$ state reached in the photoeffect is a particularly favorable case for the use of h.h. The crucial question is whether convergence for scattering problems will be similar to that found for the trinucleon bound state. In summary, it seems that future progress in the enormous field of calculations on nucleon-deuteron (elastic and inelastic) scattering, with polarization effects will involve some of the newly developed methods of solution of the Faddeev equations, applied to a local (or in general non-separable) potential.

On the other hand, the reasonable speed of convergence found for the hyperspherical expansion applied to the alpha particle, together with the meager progress by other calculational techniques, suggests that the mass-4 system, at least, is best solved by this expansion. The outlook for solution of heavier nuclei is unclear: will it involve further developments with the optimal subset for hyperspherical expansions; or treatments involving three (or four) subsystems, with multi-body forces; or use of present or modified Brueckner-type calculations? (Note that one great advantage of reliable calculations on the three and four nucleon systems by other techniques is that these allow checks on the accuracy of approximations made in Brueckner-type calculations. The comparison in Section 6 suggests that the accuracy is surprisingly good for the ground state of the triton.)

7.8 The Gordian Knot Again

The above highly subjective discussion dealt entirely with the problem of calculations for a model *assuming* a given nucleon-nucleon potential. We must return briefly to a consideration of problems of the Gordian knot, whose loose strands have appeared frequently throughout my discussion. Can we cut the Gordian knot by the sharp sword of reliable dispersion theory calculations of the nucleon-nucleon potential, of three-body forces, and of non-additive effects in electron-deuteron and electron-trinucleon scattering? Recent progress gives us some hope of success. But nevertheless I shall spend a little while on a more phenomenological approach, involving picking at the knot.

Consider first the nucleon electromagnetic form factors and non-additive effects, needed in the analysis of electron-deuteron and electron-trinucleon scattering. I emphasize the analysis of the electric (rather than the magnetic) deuteron and trinucleon form factors, to minimize the importance of non-additive and relativistic effects. The remaining non-additive effects could be treated in an iterative manner, starting with a model for the wavefunctions, and for the magnetic contribution of non-additive effects. The wavefunctions would be checked by electric form factors, and the non-additive effects by the magnetic form factors. Improved approximations to the wavefunction would be used to obtain more accurate non-additive magnetic effects; while improved models for non-additive effects would be used to correct the wavefunction found from the electric form factors.

The above analysis needs two further inputs: 1. the electric neutron form factor G_{En}; and 2. separation of the monopole and quadrupole form factors $G_0(q)$ and $G_2(q)$ of the deuteron by polarization measurements. The former is hard to measure directly; but can be found theoretically using a model in which the isoscalar spectral function is dominated by poles at the known positions of the two isoscalar meson resonances (ω, and ϕ). (Levinger, 1966; Iachello et al., 1973). The experiments on deuteron polarization effects should provide significant information on the deuteron wave function, and related off-shell t-matrix.

Our knowledge of the spin-triplet t-matrix is inaccurate even regarding on-shell values (particularly the mixing parameter $\bar{\varepsilon}_1$). Knowledge of the off-shell behavior would be substantially increased by measurements of $G_0(q)$ and $G_2(q)$ separately. These would determine the existence and size of a "hole" in the deuteron wavefunction; and also the poorly known percentage of D-state. In principle, we could then limit ourselves to unitary transformations for which the spin-triplet potential was both phase-equivalent and eigenfunction equivalent. But the absence of a bound state of the singlet S system means that we would have the greater freedom of only phase equivalence for this case. The condition that the potential should agree with OPEP at large distance provides a constraint in both spin states. Present calculations disagree as to the changes in trinucleon energy due to specified unitary transformations. Also, we must establish criteria as to which unitary transformations are "reasonable" and which are "pathological". The determination of the off-shell behavior of the t-matrix seems to be the most serious tangle in the Gordian knot.

I pointed out in Chapter I that the absence of a free neutron target posed a serious problem in the determination of nucleon-nucleon forces, unless we were willing to take the simple way out of *assuming* charge independence and charge symmetry. It is now clear from Fabre's model-insensitive calculation of the Coulomb energy of the mass-3 isodoublet that there is a small failure of charge symmetry, plausibly ascribed to isospin mixing of meson resonances exchanged between the two nucleons. The Coulomb energy discrepancy gives us one number, which can be interpreted to give a more attractive $n - n$ than $p - p$ potential, of the same effective range. Alternatively, one could say that the range is different; but the theoretical interpretation favors the change of strength. The accuracy of the assumption of charge independence is not settled: it is clear that corrections are needed for the mass difference between neutral and charged pions; but it is unclear if other corrections should be made.

Returning to the trinucleon problem: we found in Chapter VI a tangle of four effects: 1. the assumption for the non-relativistic off-shell t-matrix; 2. relativistic effects in the two and three nucleon problems; 3. three-body forces; 4. new non-additive effects in electron-trinucleon scattering. Perhaps we shall be able to disentangle these four strands by the following procedure. First, *extend* calculations of the relativistic *kinematic* effects; and lump the relativistic effects on the t-matrix with other uncertainties in finding the off-shell t-matrix. Second, treat non-additive effects in electron scattering by the iterative procedure proposed above for electron-deuteron scattering. Third, interpret any "remaining effects" in the triton as due to three-body forces.

The joker in the last sentence is the phrase "remaining effects". The procedure outlined leaves considerable uncertainty in the off-shell t-matrix for the nucleon-nucleon system. So one choice t_1 for extrapolation would lead to a choice $V_1^{(3)}$ for the three-body force needed to match triton properties; another choice t_2 would lead to $V_2^{(3)}$; etc. Noyes (1972 B) argued that this tangle of extrapolation of the $N - N$ t-matrix and the three-body force could *never* be resolved; and that the properties of the triton could be calculated to fit experiment using just a suitably chosen extrapolation of the $N - N$ t-matrix. (The above is my re-phrasing of Noyes' argument.) Brayshaw (1973 B, 1973 C) has provided a calculational formalism to combine the tangle of these two unknowns in a single quantity, the boundary condition at small distance in the three-body system. He finds that even continuum states of the three-body system provide only the value of this single parameter.

However, the extrapolation of the $N-N$ t-matrix is not completely arbitrary: we have the constraint that the potential should be local and agree with OPEP at large separation; and we hope to have the constraint on the deuteron wavefunction from improved measurements of electron-deuteron scattering. Besides these limitations, physicists always have the right to limit themselves to "reasonable extrapolations", as defined by themselves. In any treatment of experimental data one is forced to impose the condition of "reasonableness" on an extrapolation, or even on an interpolation: otherwise one could have a very narrow resonance between any two measured points for the on-shell t-matrix. We discard such unobserved resonances on the basis of reasonableness, or equivalently, using Occam's razor. I hope that the above objective and subjective constraints on extrapolation of the $N-N$ t-matrix will serve to disentangle these two strands.

A further hope of sorting out this tangle lies in Brayshaw's (1973A) analysis of electron-trinucleon elastic scattering. While I think it is premature to assert that this analysis establishes the existence of three-body forces, it certainly provides an additional loose strand that is likely to help in picking at the knot. The charge form factor brings in non-additive contributions; but perhaps they could in turn be disentangled by the iterative procedure suggested above for the analysis of electron-deuteron scattering.

Perhaps remaining tangles as to what is due to the off-shell extrapolation of the t-matrix and what is due to three-body forces could be sorted out from calculations on the energy of the alpha particle, where one could reasonably hope that these two unknowns would enter in a different manner than in the trinucleon. Of course, we would need to calculate corrections for relativistic kinematics in the alpha particle. But now we face the additional problem of introducing *four-body forces*. It seems to be a general rule in phenomenological nuclear physics that whenever we study a new experimental quantity to sort out unknown effects, we introduce still another unknown.

Acknowledgement. The first draft of this manuscript was written as lecture notes for a "Cours de recherche" which I gave at the Institut de Physique Nucléaire, Orsay, in the fall of 1972, while I was on sabbatical leave from Rensselaer Polytechnic Institute. I wish to express my gratitude to the I.P.N. and to R.P.I., and also to the Fulbright Commission for a travel grant. I also want to thank the physicists of the I.P.N. for their hospitality and for many helpful comments on my lectures and lecture notes. I also received very helpful detailed comments from about ten other nuclear physicists.

References

Aaron,R., Amado,R.D., Yam,Y.Y.: Phys. Rev. **140**B, 1291 (1965). Section 7.4.

Aaron,R., Amado,R.D.: Phys. Rev. **150**, 857 (1966). Section 7.5.

Aarons,J.C., Sloan,I.H.: Nucl. Phys. A**182**, 369 (1972). Sections 7.4 and 7.5.

Adler,R.K.: Phys. Rev. **141**, 1499 (1966). Sections 2.3 and 2.4.

Afnan,I.R., Serduke,F.J.D.: Phys. Letters **44**B, 143 (1973). Sections 4.2 and 6.9.

Afnan,I.R., Read,J.M.: Australian J. Phys. **26**, (1973A). Sections 4.2, 4.4 and 6.2.

Afnan,I.R., Read,J.M.: Phys. Rev. C**8**, 1294 (1973B). Sections 5.6 and 6.2; Table 6.2; Fig. 6.1.

Ahmadzadeh,A., Tjon,J.A.: Phys. Rev. **139**B, 1085 (1965). Sections 5.3 and 6.3.

Alfimenkov,V.P., Lushchikov,V.I., Kikolenko,V.G., Taran,Yu.V.L., Shapiro,F.L.: Phys. Letters **24**B, 151 (1967). Section 7.2.

Allard,J.C., Armstrong,A.H., Rosen,L.: Phys. Rev. **91**, 90 (1953). Section 7.4.

Alt,E.O., Grassberger,P., Sandhas,W.: Nucl. Phys. B**2**, 167 (1967). Sections 6.2 and 7.5.

Amado,F.: Ann. Rev. Nucl. Sci. **19**, 61 (1969). Sections 1.1, 7.3, and 7.4; Figs. 7.3, 7.4 and 7.5.

Amado,R.D.: Phys. Rev. C**2**, 2439 (1970). Section 3.6.

Amado,R.D.: In Los Angeles conference, "Few Particle Problems", Ed.: I. Slaus et al., p. 254. New York: American Elsevier 1972. Section 7.1.

Arenhövel,H., Weber,H.J.: Springer Tracts Mod. Phys. **65**, 58 (1972). Sections 2.3 and 2.4.

Avishai,Y., Rinat-Reiner,A.S.: Phys. Letters **37**B, 487 (1971). Section 7.4.

Badalyan,A.M.: Yadern. Fiz. **8**, 313 (1968), Soviet J. Nucl. Phys. (transl.) **8**, 180 (1969). Section 7.1.

Bahethi,O.P., Fuda,M.G.: Phys. Rev. C**6**, 1956 (1972). Section 6.9.

Baker,G.H.,Jr.: Advan. Theoret. Phys. **1**, 1 (1965). Section 6.3.

Ball,J.S., Wong,D.Y.: Phys. Rev. **169**, 1362 (1968). Section 4.3.

Ballot,J.L., Beiner,M., Fabre de la Ripelle,M.: Proc. of Symposium on Present Status and Novel Developments in Many-Body Problem (Rome, Sept. 1972). Sections 6.10 and 7.5.

Baranger,M., Giraud,B., Mukhopadyyay,S.K., Sauer,P.U.: Nucl. Phys. A**138**, 1 (1969). Sections 1.4, 3.6 and 4.4.

Barbour,I.M., Phillips,A.C.: Phys. Rev. C**1**, 165 (1970). Sections 7.2 and 7.5; Fig. 7.7.

Barschall,H.H., Haeberli,W. (Eds.): "Polarization Phenomena in Nuclear Reactions", pp. 25–29. Madison: Univ. Wisconsin Press 1971. Section 2.4.

Bartolini,W., Donaldson,R.E., Graves,D.J.: Phys. Rev. **174**, 313 (1968). Section 7.2.

Barton,G., Phillips,A.C.: Nucl. Phys. A**123**, 97 (1969). Section 7.3.

Basdevant,J.L.: Fortschr. Physik **20**, 283 (1972). Section 6.3.

Beiner,M., Fabre de la Ripelle,M.: Letters Nuovo Cimento **1**, 584 (1971). Section 6.5.

Beiner,M., Gara,P.: Comp. Phys. Commun. **4**, 1 (1971). Section 6.5.

Benayoun,J.J., Gignoux,C.: Nucl. Phys. A**190**, 419 (1972). Section 6.3.

Berick,A.C., Riddle,R.A., York,C.M.: Phys. Rev. **174**, 1105 (1968). Section 7.4.

Bernheim,M., Blum,D., McGill,W., Rishalla,R., Trail,C., Stovall,T., Vinciguerra,D.: Letters Nuovo Cimento **5**, 431 (1972). Section 6.9, and Fig. 6.7.

Bhakar,B.S., Mitra,A.N.: Phys. Rev. Letters **14**, 143 (1965). Section 5.5.

Bhatia,R.P., Walker,J.F.: Nucl. Phys. A**192**, 658 (1972). Section 3.7.

Bhatt,S.C., Harms,E., Levinger,J.S.: Phys. Letters **40**B, 23 (1972). Sections 4.2, 5.5 and 5.6; Tables 4.1 and 6.2.

Bhatt,S.C.: Can. J. Phys. **51** (1973). Section 4.2.

Biedenharn,L.C., Blatt,J.M., Kalos,M.H.: Nucl. Phys. **6**, 359 (1958). Section 2.3.

Blatt,J.M., Weisskopf,V.F.: „Theoretical Nuclear Physics". New York: John Wiley and Sons, 1952. Sections 2.3, 3.3, 3.4, 3.5, 3.6, 6.4, 6.8 and 6.9.

Blatt,J.M., Biedenharn,L.C.: Phys. Rev. **86**, 399 (1952). Section 4.2.

Bodmer,A.R.: Phys. Rev. **141**, 1387 (1966). Section 7.6.

Bolsterli, M., Mackenzie, J.: Physics **2**, 141 (1965). Sections 3.6 and 3.7.

Bolsterli, M.: Phys. Rev. **182**, 1095 (1969). Section 3.5.

Brady, T.: Ph. D. dissertation, Rensselaer Polytechnic Institute, August, 1969. Section 2.4.

Brady, T., Fuda, M., Harms, E., Levinger, J. S., Stagat, R.: Phys. Rev. **186**, 1069 (1969). Sections 2.3, 3.5, 3.6, 4.2, 5.4, 5.5 and 5.6; Fig. 5.3.

Brady, T.: Phys. Letters **32** B, 85 (1970). Sections 4.3, 5.6, 6.2 and 6.6.

Brady, T., Harms, E., Laroze, L., Levinger, J. S.: Phys. Rev. C **2**, 59 (1970). Section 7.3; Fig. 7.1.

Brady, T., Harms, E., Laroze, L., Levinger, J. S.: Nucl. Phys. A **168**, 507 (1971). Section 6.4.

Brady, T., Levinger, J. S., Tomusiak, E. L.: Bull. Am. Phys. Soc. **17**, 438 (1972). Sections 2.3, 2.4 and 4.2. Figures 2.7, 2.8, 2.9, 2.10 and 2.11.

Brady, T., Sloan, I. H.: Phys. Letters **40** B, 55 (1972). Sections 4.3 and 7.5.

Brady, T., Sloan, I. H.: Phys. Rev. C **8**, (1973). Section 7.5.

Brayshaw, D. D., Buck, B.: Phys. Rev. Letters **24**, 733 (1970). Section 3.6.

Brayshaw, D. D.: Phys. Rev. D **7**, 1835 (1973 A). Sections 6.10 and 7.8.

Brayshaw, D. D.: Phys. Rev. D **8**, 952 (1973 B). Sections 7.5 and 7.8.

Brayshaw, D. D.: Bull. Am. Phys. Soc. **18**, 1391 (1973 C). Sections 7.5 and 7.8.

Breit, G., Harracz, R. D.: In: Burhop, E. H. (Ed.): High Energy Physics, Vol. 1, p. 21. New York: Academic Press, 1967. Section 2.1.

Brown, G. E. in Los Angeles conference, "Few Particle Problems, (Ed.): Slaus, I. et al., p. 135. New York: American Elsevier, 1972. Section 6.9.

Bruinsma, J. R., Wageninger, R. van: Phys. Letters **44** B, 221 (1973). Section 6.5 and Table 6.2.

Burg, J. P., Cabrillat, J. C., Chemarin, M., Ille, B., Nicolai, G.: Budapest Conference, July (1971), Abstract C–10. Section 7.5.

Burnap, C., Levinger, J. S., Siebert, B.: Phys. Letters **33** B, 337 (1970). Sections 2.3 and 4.2.

Cahill, R. T., Sloan, I. H.: Nucl. Phys. A **165**, 161 (1971). Section 7.5.

Cahill, R. T., Sloan, I. H.: Nucl. Phys. A **194**, 589 (1972). Section 7.5.

Catron, H. C., Goldberg, M. D., Hill, R. W., LeBlanc, J. M., Stoering, J. P., Taylor, C. J., Williamson, M. A.: Phys. Rev. **123**, 218 (1961). Section 7.4.

Chemtob, M., Durso, J. W., Riska, D. O.: Nucl. Phys. B **38**, 141 (1972). Section 2.2.

Collard, H., Hofstadter, R., Hughes, E. B., Johansson, A., Yearian, M. R.: Phys. Rev. **138** B, 57 (1965). Section 6.10; Table 6.4; Fig. 6.7.

Cottingham, W. N., LaCombe, M., Loiseau, B., Richard, J. M., Vinh Mau, R.: Phys. Rev. D **8**, 800 (1973). Section 2.2.

Coury, F. M., Frank, W. M.: Nucl. Phys. **46**, 257 (1963). Section 6.9; Fig. 6.6.

Dabrowski, J., Dworzecki, M., Trych, E.: Acta Phys. Polon. **33**, 831 (1968). Section 7.6.

Dabrowski, J., Fedorynska, E.: Nucl. Phys. A **210**, 509 (1973). Section 7.6.

Delves, L. M.: Nucl. Phys. **9**, 391 (1959). Section 6.5.

Delves, L. M.: Nucl. Phys. **20**, 275 (1960). Section 6.5.

Delves, L. M., Phillips, A. C.: Rev. Mod. Phys. **41**, 497 (1969). Sections 1.1, 6.4, 7.2 and 7.3.

Delves, L. M., Hennell, M.: Nucl. Phys. A **168**, 347 (1971). Sections 6.4, 6.9, 7.3, 7.5 and 7.6.

Delves, L. M.: Advan. Nucl. Phys. **5**, 1 (1972). Sections 1.1 and 6.4; Table 6.3; Fig. 6.2.

Demin, V. F., Efros, V. D.: JETP Letters (transl.) **16**, 360 (1972). Section 7.6.

Demin, V. F., Pokrovsky, Yu. E., Efros, V. D.: Phys. Letters **44** B, 227 (1973). Sections 6.5 and 7.6; Tables 6.2, 6.4.

Dilg, W., Koester, L., Nistler, W.: Phys. Letters **36** B, 208 (1971). Section 7.2.

Doleschall, P.: Phys. Letters **38** B, 298 (1972). Section 7.4.

Doleschall, P., Aarons, J. C., Sloan, I. H.: Phys. Letters **40** B, 605 (1972). Section 7.4.

Doleschall, P.: Nucl. Phys. A **201**, 264 (1973). Section 7.4.

Downs, B. W.: Nuovo Cimento **43** A, 454 (1966). Section 7.6.

Duck, Ian: Advan. Nucl. Phys. **1**, 343 (1968). Section 7.3.

Ebenhöh,W.: Nucl. Phys. A **191**, 97 (1972). Section 7.5; Fig. 7.6.

Efimov,V.: Phys. Letters **33**B, 563 (1970). Section 7.1.

Efros,V.D.: Yad. Fiz. **15**, 226 (1972). Section 6.5.

Eikemeier,H., Hackenbroich,H.H.: Nucl. Phys. A **169**, 407 (1971). Sections 6.5 and 7.6.

Elias,J.E., Friedman,J.I., Hartmann,G.C., Kendall,H.W., Kirk,P.N., Sogard,M.R., Speybroeck,L.P.van: Phys. Rev. **177**, 2075 (1969). Section 2.4.

Erens,G., Visschers,J.L., Wageningen,R.van: Ann. Phys. (N.Y.) **67**, 461 (1971). Sections 6.4 and 6.5; Table 6.1; Fig. 6.4.

Ernst,D.J., Shakin,C.M., Thaler,R.M.: Phys. Rev. C **8**, 46 (1973), and Phys. Rev. C **8**, 2056 (1973). Sections 4.1 and 4.3.

Fabre de la Ripelle,M.: Rev. Roumaine Phys. **14**, 1215 (1969). Section 6.5.

Fabre de la Ripelle,M.: C. R. Acad. Sci. Paris **273**, 1007 (1971). Section 6.5.

Fabre de la Ripelle,M.: Fiz. **4**, 1 (1972). Sections 2.5, 6.1 and 6.8.

Fabre de la Ripelle,M., Levinger,J.S.: Bull. Am. Phys. Soc. **18**, 1392 (1973). Section 7.5; Fig. 7.7.

Faddeev,L.D.: Zh. Eksperim. Teor. Fiz. **39**, 1459 (1960); Sov. Phys. JETP (transl.) **12**, 1014 (1961). Section 5.1.

Fermi,E., Marshall,K.: Phys. Rev. **75**, 578 (1947). Section 7.3.

Feshbach,H.: Phys. Rev. **107**, 1626 (1957). Section 2.3.

Feshbach,H., Kerman,A.K.: Comments Nucl. Part. Phys. **1**, 132 (1967); **2**, 22 (1968); **2**, 78 (1968). Section 2.2.

Fetisov,V.N., Gorbunov,A.N., Varfolomeev,V.T.: Nucl. Phys. **91**, 305 (1965). Section 7.5; Fig. 7.7.

Fiedeldey,H.: Nucl. Phys. A **135**, 353 (1969). Sections 2.3, 4.1, 4.3 and 6.9.

Fiedeldey,H., McGurk,N.J.: Nucl. Phys. A **189**, 83 (1972). Section 6.9.

Fiore,M., Arvieux,J., Sen,N.van, Perrin,G., Merchez,F., Gardraud,J.C., Perrin,C., Durand,J.L., Darves-Blanc,R.: Phys. Rev. C **8**, 2019. (1973). Section 7.4.

Fuda,M.G.: Nucl. Phys. A **116**, 83 (1968A). Sections 3.1, 3.3, 3.7, 4.1, and 5.5.

Fuda,M.G.: Phys. Rev. **166**, 1064 (1968B). Sections 3.5 and 6.2.

Fuda,M.G.: Nucl. Phys. A **130**, 155 (1969). Section 4.4.

Fuda,M.G.: Phys. Rev. C **1**, 1910 (1970). Sections 4.2 and 4.4.

Fujita,J., Miyazawa,H.: Prog. Theor. Phys. (Kyoto) **17**, 360 (1957). Section 6.9; Fig. 6.6.

Galster,G., Klein,H., Moritz,J., Schmidt,K.H., Wegener,D., Bleckwenn,J.: Nucl. Phys. B **32**, 221 (1971). Section 2.4.

Gignoux,C., Laverne,A.: Phys. Rev. Letters **29**, 436 (1972). Sections 6.3, 6.10 and 7.3; Fig. 6.7.

Gissler,W.: Z. Krist. **118**, 149 (1963). Section 7.2.

Glendening,N.K., Kramer,G.: Phys. Rev. **126**, 2159 (1962). Sections 2.1 and 2.4.

Gogny,D., Pires,P., Tourreil,R.de: Phys. Letters **32**B, 591 (1970). Sections 6.5, 7.5 and 7.6.

Goldberger,M.L., Watson,K.M.: Collision Theory. New York: Wiley 1964. Sections 3.1, 3.2 and 3.6.

Gourdin,M., Piketty,C.A.: Nuovo Cimento **32**, 1137 (1964). Section 2.4.

Gourdin,M.: Diffusion des Electrons de Hautes Energie. Paris: Masson et Cie. 1966. Sections 1.2, 1.4, 2.1 and 2.4.

Gourdin,M.: Proc. Daresbury Study Weekend, No. 1, Editors A. Donnachi and E. Gabathuler (Science Research Council, Daresbury, 1970), 95. Section 2.5.

Grashin,A.F.: J. Exptl. Theor. Phys. (USSR) **36**, 1717 (1959); Sov. Phys. JETP (transl.) **9**, 1223 (1959). Section 2.2.

Gupta,V.K., Mitra,A.N.: Phys. Rev. Letters **15**, 974 (1965). Section 6.9.

Gupta,V.K., Bhakar,B.S., Mitra,A.N.: Phys. Rev. **153**, 1114 (1967). Sections 3.6 and 5.4.

Hadjimichael,E., Jackson,A.D.: Nucl. Phys. A **180**, 217 (1972). Sections 6.4, 6.9 and 6.10; Tables 6.2 and 6.4.

Haftel, M. I.: Phys. Rev. Letters **25**, 120 (1970). Section 3.6.

Haftel, M. I., Tabakin, F.: Phys. Rev. C **3**, 921 (1971). Sections 3.6 and 6.9.

Haftel, M. I.: Phys. Rev. C **7**, 80 (1973). Sections 2.3, 2.4 and 3.6; Figs. 2.4, 2.13, 6.3 and 6.9.

Hamada, T., Johnston, I. D.: Nucl. Phys. **34**, 382 (1962). Section 2.3; Fig. 2.2.

Harms, E.: Ph. D. Dissertation, Rensselaer Polytechnic Institute, August 1969. Section 3.7.

Harms, E., Levinger, J. S.: Phys. Letters **30** B, 449 (1969). Section 5.4; Table 6.1.

Harms, E.: Phys. Rev. C **1**, 1667 (1970 A). Sections 3.1, 3.7, 4.1, 4.3, 5.4 and 6.2; Tables 4.3 and 4.4; Figs. 3.2, 3.3, 3.4, 3.5, 3.8, 3.9, 3.10, 3.11 and 3.12.

Harms, E.: Nucl. Phys. A **159**, 545 (1970 B). Section 5.6.

Harms, E., Laroze, L.: Nucl. Phys. A **160**, 449 (1970). Sections 3.7, 4.3, 5.4 and 5.6; Tables 5.1 and 6.1; Figs. 3.6, 3.7 and 5.4.

Harms, E.: XIV Latin American School of Physics, Caracas, July, 1972 A, Reidel, Holland. Sections 1.1 and 7.3.

Harms, E.: Phys. Letters **41** B, 26 (1972 B). Section 7.5.

Harper, E. P., Kim, Y. E., Tubis, A.: Phys. Rev. C **2**, 877 (1970) and Phys. Rev. C **2**, 2455 E (1970). Section 6.3.

Harper, E. P., Kim, Y. E., Tubis, A.: Phys. Rev. C **6**, 126 (1972 A). Sections 5.3 and 6.3.

Harper, E. P., Kim, Y. E., Tubis, A.: Phys. Rev. Letters **28**, 1533 (1972 B). Section 6.3; Table 6.4.

Harper, E. P., Kim, Y. E., Tubis, A.: Phys. Rev. C **6**, 1601 (1972 C). Section 6.9.

Harrington, D.: Phys. Rev. **147**, 685 (1966). Section 7.6.

Hartt, K., Sullivan, E.: Phys. Rev. D **4**, 1353 (1971). Section 7.6.

Henley, E. M., Kellher, T. E.: Nucl. Phys. A **189**, 632 (1972). Section 2.5; Fig. 2.14.

Henley, E. M.: In: Los Angeles Conference, "Few Particle Problems", Editors I. Slaus et al., New York: American Elsevier, 1972, p.221. Section 2.5.

Hennell, M. A., Delves, L. M.: Phys. Letters **40** B, 20 (1972). Sections 6.4, 6.9 and 7.6; Table 6.2.

Hennell, M. A., Delves, L. M.: Acta Phys. Hung. **33**, 103 (1973 A). Section 6.4; Tables 6.2 and 6.4; Fig. 6.3.

Hennell, M. A., Delves, L. M.: Nucl. Phys. A **204**, 552 (1973 B). Section 6.9.

Herndon, R. C., Tang, Y. C.: Phys. Rev. **159**, 853 (1967). Section 7.6.

Hetherington, J. H., Schick, L. H.: Phys. Rev. **137** B, 935 (1965). Sections 7.4 and 7.6.

Hoffman, R. A., Lefrancois, J., Thorndike, E. H., Wilson, R.: Phys. Rev. **125**, 973 (1962). Section 7.2.

Howerton, R. J.: Tabulated Neutron Cross Sections, Univ. Calif. Radiation Lab. Rept. UCRL 5573 (1961, unpublished). Section 7.4.

Humbertson, J., Hall, R. L., Osborn, T. A.: Phys. Letters **27** B, 195 (1968). Sections 6.3, 6.4 and 7.1.

Iachello, F., Jackson, A. D., Lande, A.: Phys. Letters **43** B, 191 (1973). Section 1.2.

Jackson, A. D., Tjon, J. A.: Phys. Letters **32** B, 9 (1970). Section 6.9.

Jackson, A. D., Lande, A., Sauer, P. U.: Nucl. Phys. A **156**, 1 (1970). Section 6.4.

Jackson, A. D., Lande, A., Sauer, P. U.: Phys. Letters **35** B, 365 (1971). Section 6.4; Table 6.2.

Kharchenko, V. F.: Ukr. Phys. J. **7**, 573 (1962) and **7**, 582 (1962). Section 5.4.

Kharchenko, V. F., Petrov, N. M., Storozhenko, S. A.: Nucl. Phys. A **106**, 464 (1968). Section 5.5.

Kim, Y. E.: J. Math. Phys. **10**, 1491 (1969). Section 6.3.

Kim, Y. E.: Book on trinucleon, to be published by Oxford Univ. Press, 1974. Section 6.3.

Kloet, W. M., Tjon, J. A.: Phys. Letters **37** B, 460 (1971). Section 7.5.

Kloet, W. M., Tjon, J. A.: Ann. Phys. (N.Y.) **79**, 407 (1973). Section 7.5.

Kok, L. P., Erens, G., Wageningen, R. van: Nucl. Phys. A **122**, 684 (1968). Section 5.4.

Kok, L. P.: Thesis, University of Amsterdam, 1970. Section 4.3.

Kowalski, K. L.: Phys. Rev. Letters **15**, 538 (1965). Section 3.7.

Kowalski, K. L., Monahan, J. E., Shakin, C. M., Thaler, R. M.: Phys. Rev. C 3, 1146 (1971). Section 3.6.

Lambacher, H., Urban, P.: Acta Phys. Austr. 36, 39 (1972). Section 4.2.

Laroze, L.: Ph. D. Dissertation, Rensselaer Polytechnic Institute, Jan. 1970. Sections 6.9 and 7.3.

Laroze, L., Harms, E., Levinger, J. S.: Nucl. Phys. A 158, 615 (1970). Section 7.4.

Lasinski, T. A., Barbaro-Galtieri, A., Kelly, R. L., Rittenberg, A., Rosenfeld, A. H., Trippe, T. G., Barash-Schmidt, N., Bricman, C., Chaloupka, V., Söding, P., Roos, M.: Rev. Mod. Phys. 45, 51 (1973). Sections 2.2 and 2.5.

Laverne, A., Gignoux, C.: Nucl. Phys. A 203, 597 (1973). Sections 6.3 and 6.9; Tables 6.2 and 6.4.

Lavine, J. P., Mukhopadhyay, S. K., Stephenson, G. J., Jr.: Phys. Rev. C 7, 968 (1973). Section 3.5.

Lehman, D. R.: Phys. Rev. Letters 23, 1339 (1969). Section 7.2.

Levinger, J. S., Rojo, O.: Phys. Rev. 123, 2177 (1961). Sections 2.3 and 2.4.

Levinger, J. S.: Phys. Letters 29 B, 216 (1969). Section 2.3.

Levinger, J. S., Lu, A. H., Stagat, R.: Phys. Rev. 179, 926 (1969). Section 3.6.

Levinger, J. S., O'Donoghue, J.: Bull. Am. Phys. Soc. 17, 440 (1972). Section 3.7; Fig. 3.13.

Levinger, J. S.: Acta Phys. Hung. 33, 135 (1973 A). Sections 1.1, 1.3, 1.4, 2.4, 3.6 and 3.7; Fig. 3.1.

Levinger, J. S.: J. Math. Phys. 14, 1314 (1973 B). Section 3.5.

Lomon, E.: Comm. Nucl. Part. Phys. 4, 28 (1970). Section 2.2.

Louck, J. D., Galbraith, H. W.: Rev. Mod. Phys. 44, 540 (1972). Section 6.5.

Lovelace, C.: Phys. Rev. 135 B, 1225 (1964). Sections 3.1, 3.3, 3.7 and 5.2.

MacGregor, M. H., Arndt, R. A., Wright, R. M.: Phys. Rev. 182, 1714 (1969). Sections 2.1 and 4.2.

Malfliet, R. A.: Ph. D. Thesis, University of Utrecht (1969). Section 6.3; Table 6.1; Fig. 6.7.

Malfliet, R. A., Tjon, J. A.: Nucl. Phys. A 127, 161 (1969). Sections 5.4, 6.1, 6.2 and 6.3; Table 6.1.

Malfliet, R. A., Tjon, J. A.: Ann. Phys. (N.Y.) 61, 425 (1970). Sections 6.3 and 7.3.

Malfliet, R. A., Tjon, J. A.: In: Los Angeles Conference, "Few Particle Problems", Editors I. Slaus et al., p. 441. New York: American Elsevier, 1972. Section 6.3; Tables 6.2 and 6.4.

McCarthy, J. S., Sick, I., Whitney, R., Yearian, M. R.: Phys. Rev. Letters, 25, 884 (1970). Section 6.10.

McKee, J. S. C., Conzett, H. E., Larimer, R. M., Leeman, Ch.: Phys. Rev. Letters 29, 1613 (1972). Section 7.2.

Messiah, A.: Quantum Mechanics. New York: Interscience 1961. Sections 3.1 and 3.2.

Mitra, A. N., Narasimham, V. L.: Nucl. Phys. 14, 407 (1959). Section 4.1.

Mitra, A. N., Naqvi, J. H.: Nucl. Phys. 25, 307 (1961). Section 3.5.

Mitra, A. N.: Nucl. Phys. 32, 529 (1962). Sections 3.5, 5.1 and 5.4.

Mitra, A. N.: Advan. Nucl. Phys. 3, 1 (1969). Sections 1.1, 5.1 and 7.3.

Mongan, T. R.: Phys. Rev. 175, 1260 (1968). Sections 4.1, 4.3 and 4.4.

Mongan, T. R.: Phys. Rev. 178, 1597 (1969). Sections 2.3, 2.4, 2.6, 4.1, 4.3, 4.4 and 6.6.

Moravcsik, M. J.: The Two-Nucleon Interaction. London: Oxford Univ. Press, 1963. Sections 2.1 and 2.2.

Moravcsik, M. J.: Rep. Prog. Phys. 35, 587 (1972). Section 2.1.

Mukherjee, S., Shyam, R.: Phys. Rev. C 8, 1149 (1973). Section 2.3.

Nissimov, H.: Phys. Letters 46 B, 1 (1973). Section 6.4.

Noyes, H. P.: Phys. Rev. Letters 15, 538 (1965). Section 3.7.

Noyes, H. P.: Prog. Nucl. Phys. 10, 355 (1969 A). Sections 2.1, 2.2 and 2.3.

Noyes, H. P.: Phys. Rev. Letters 23, 1201 (1969 B); Phys. Rev. Letters 24, 493 (1970) E. Section 7.5.

Noyes, H. P.: Proc. of 1st Intl. Conf. 3-body problem, 1970, p. 2. Section 6.5.
Noyes, H. P.: Ann. Rev. Nucl. Sci. **22**, 465 (1972A). Sections 2.1 and 6.9.
Noyes, H. P.: In Los Angeles Conference, "Few Particle Problems", p. 122. Ed.: Slaus, I. et al. New York: American Elsevier, 1972B. Sections 3.6 and 7.8.
O'Connell, J. S., Prats, F.: Phys. Rev. **184**, 1007 (1969). Section 7.5.
Ohlsen, G. G.: Rep. Prog. Phys. **35**, 717 (1972). Section 2.4.
Okamoto, K.: Phys. Letters **11**, 150 (1964), Prog. Theor. Phys. (Kyoto) **34**, 326 (1965). Section 6.8.
Okamoto, K., Pask, C.: Ann. Phys. (N.Y.) **68**, 18 (1971). Section 6.8.
Osborn, T. A.: Ph. D. Thesis, Stanford Univ. 1967; SLAC-PUB-361 (1967). Section 6.3.
Osborn, T. A.: Nucl. Phys. A **138**, 305 (1969). Section 4.3.
Osborn, T. A.: J. Math. Phys. **14**, 373 (1973) and **14**, 1485 (1973). Section 3.5.
Panchapakesan, N.: Phys. Rev. **140**, 1320 (1965). Section 7.6.
Partovi, F., Lomon, E. L.: Phys. Rev. D **5**, 1192 (1972). Section 2.3.
Pask, C.: Phys. Letters **25**, 78 (1967). Section 6.9.
Phillips, A. C.: Phys. Rev. **142**, 984 (1966). Section 6.9.
Phillips, A. C.: Nucl. Phys. A **107**, 209 (1968). Sections 2.3, 3.6, 4.2, 5.5 and 7.3; Fig. 5.2.
Picker, H. S., Redish, E. F., Stephenson, G. J., Jr.: Phys. Rev. C **4**, 287 (1971). Section 3.6.
Pieper, S. C., Kowalski, K. L.: Phys. Rev. C **5**, 306 (1972).
Pieper, S. C.: Nucl. Phys. A **193**, 519 (1972A). Section 7.5.
Pieper, S. C.: Nucl. Phys. A **193**, 529 (1972B). Sections 7.4 and 7.5.
Pieper, S. C.: Phys. Rev. C **8**, 1702 (1973). Section 7.41.
Pniewski, J.: In: Los Angeles Conference, "Few Particle Problems", p. 145. Editors I. Slaus et al. New York: American Elsevier, 1972. Section 7.6.
Razavy, M., Field, G., Levinger, J. S.: Phys. Rev. **125**, 269 (1962). Sections 2.3 and 3.4.
Reid, R. V.: Ann. Phys. (N.Y.) **50**, 411 (1968). Sections 2.3, 2.6, 3.1, 6.1 and 6.2; Fig. 2.2.
Reid, R. V., Vaida, M. L.: Phys. Rev. Letters **8**, 494 (1972). Section 2.1.
Riihmäki: Ph. D. Thesis, Massachusetts Institute of Technology (1970). Section 2.6.
Sachs, R. G.: Nuclear Theory. Cambridge, Mass.: Addison-Wesley, 1953. Section 1.5.
Sauer, P. U.: Ann. Phys. (N.Y.) **80**, 242 (1973). Section 3.6, 4.4
Sauer, P. U., Tjon, J. A.: Proc. (Munich) Intl. Conf. Nucl. Phys. **1**, p. 25 (1973). New York: American Elsevier. Section 3.6 and 6.9.
Sauer, P. U., Stingl, M.: Proc. (Munich) Intl. Conf. Nucl. Phys. **1**, p. 24 (1973). New York: American Elsevier. Section 6.9.
Schiff, L. I.: Phys. Rev. **133** B, 802 (1964), Section 6.10.
Seagrave, J. D.: Phys. Rev. **97**, 757 (1954). Section 7.4.
Seagrave, J. D., Cranberg, L.: Phys. Rev. **105**, 1816 (1957). Section 7.4.
Seagrave, J. D.: In: Three Body Problem, p. 41. Amsterdam: North Holland 1970. Figure 7.2.
Seamon, R. E., Friedman, K. A., Breit, G., Haracz, R. D., Holt, J. M., Prokash, A.: Phys. Rev. **165**, 1579 (1968). Section 2.1.
Siebert, B., Levinger J. S., Harms, E.: Nucl. Phys. A **197**, 33 (1972). Sections 4.2 and 4.4; Table 4.2; Figs. 4.1, 4.2, 4.3 and 4.4.
Siebert, B.: Ph. D. Dissertation, Rensselaer Polytechnic Institute, June, 1973, Section 4.2.
Siebert, B., Levinger, J. S.: Z. Physik, in press. (1974). Sections 4.2, 5.6 and 6.2.
Signell, P.: Advan. Nucl. Phys. **2**, 223 (1969). Section 2.1.
Signell, P.: In: Los Angeles Conference, "Few Particle Problems", p. 1. Editors I. Slaus et al. New York: American Elsevier, 1972. Sections 2.1, 2.2 and 2.3.
Simonov, Yu. A.: Yad. Fiz. **3**, 630 (1966); Sov. J. Nucl. Phys. **3**, 461 (1966). transl. Section 6.5.
Simonov, Yu. A.: Proc. of Symposium on Present Status and Novel Developments in Many Body Problem (Rome, Sept. 1972). Sections 6.5, 7.3 and 7.6.
Sitenko, A. G., Kharchenko, V. F.: Yad. Fiz. **1**, 994 (1965); Sov. J. Nucl. Phys. **1**, 708 (1965), transl. Section 5.5.

Sitenko,A.G., Kharchenko,V.F.: Usp. Fiz. Nauk. **103**, 469 (1971). Sov. Phys. Usp. **14**, 125 (1971), transl. Sections 1.1 and 7.3.
Slaus,I. et al., (Eds.): "Few Particle Problems", Los Angeles Conference. New York: American Elsevier, 1972. Section 1.1.
Sloan,I.H., Gray,J.D.: Phys. Letters **44**B, 354 (1973). Section 3.5.
Sotona,M.: Letters Nuovo Cimento **7**, 523 (1973). Sections 6.6 and 7.3.
Stagat,R.: Ph. D. Dissertation, Rensselaer Polytechnic Institute, August 1968. Section 5.4.
Stagat,R.: Nucl. Phys. A **125**, 654 (1969). Section 4.3.
Stapp,H.P., Ypsilantis,T.J., Metropolis,N.: Phys. Rev. **105**, 32 (1957). Section 4.2.
Stoler,P., Kaushal,N.N., Green,F., Harms,E., Laroze,L.: Phys. Rev. Letters **29**, 1745 (1972). Sections 7.3 and 7.4; Fig. 7.2A.
Stoler,P., Kaushal,N.N., Green,F.: Phys. Rev. C **8**, 1539 (1973). Section 7.3; Fig. 7.2B.
Tabakin,F.: Ann. Phys. (N.Y.) **30**, 51 (1964). Sections 4.1, 4.2, 4.3, 4.4, 5.6, 6.2 and 6.6.
Tabakin,F.: Phys. Rev. **137**B, 75 (1965). Sections 4.3, 5.4, 5.5, 5.6 and 5.9.
Tabakin,F.: Phys. Rev. **174**, 1208 (1968). Section 3.5.
Thomas,L.H.: Phys. Rev. **47**, 903 (1935). Section 6.4.
Tjon,J.A., Gibson,B.F., O'Connell,J.S.: Phys. Rev. Letters **25**, 540 (1970). Section 6.10.
de Tourreil,R., Sprung,D.W.L.: Nucl. Phys. A **201**, 193 (1973). Sections 2.3, 2.4 and 2.6; Fig. 2.2.
Tripathi,R.K., Goldhammer,P.: Phys. Rev. C **6**, 101 (1972). Sections 6.6, 7.3 and 7.6.
Überall,H.: Electron Scattering from Complex Nuclei. New York: Academic Press, 1971. Section 2.4.
van Dijk,W., Razavy,M.: Nucl. Phys. A **159**, 161 (1970). Section 3.6.
van Wageningen,R.: Proc. of Symposium on Present Status and Novel Developments in Many-Body Problem (Rome, Sept. 1972). Section 2.5.
Vinh Mau,R., Richard,J.M., Loiseau,B., LaCombe,M., Cottingham,W.N.: Phys. Letters **44**B, 1 (1973). Section 2.2; Fig. 2.1.
Watson,K.M., Nuttall,J.: Topics in Several Particle Dynamics. San Francisco: Holden Day, 1967. Sections 1.1, 3.1, 3.3, 5.1 and 5.2.
Weinberg,S.: Phys. Rev. **131**, 440 (1963). Section 4.3.
Williams,H.T., Arenhövel,H., Miller,H.G.: Phys. Letters **36**B, 295 (1971). Section 2.4.
Wilson,R.: The Nucleon-Nucleon Interaction. New York: Interscience 1963. Section 2.1.
Wilson,R.R., Levinger,J.S.: Ann. Rev. Nucl. Sci. **14**, 135 (1964). Sections 1.2, 2.1 and 2.4.
Yamaguchi,Y.: Phys. Rev. **95**, 1628 (1954). Sections 3.3, 3.5 and 3.7.
Yamaguchi,Y., Yamaguchi,Y.: Phys. Rev. **95**,1635 (1954). Sections 3.3, 4.2 and 5.5.
Yang,S.N., Jackson,A.D.: Phys. Letters **36**B, 1 (1971). Section 6.10.

Prof. Dr. J. S. Levinger
Rensselaer Polytechnic Institute
Troy, New York 12181, USA

SPRINGER TRACTS IN MODERN PHYSICS

Ergebnisse der exakten Naturwissenschaften

Atomic Physics

Dettmann, K.: High Energy Treatment of Atomic Collisions (Vol. 58)

Donner, W., Süßmann, G.: Paramagnetische Felder am Kernort (Vol. 37)

Racah, G.: Group Theory and Spectroscopy (Vol. 37)

Seiwert, R.: Unelastische Stöße zwischen angeregten und unangeregten Atomen (Vol. 47)

Zu Putlitz, G.: Determination of Nuclear Moments with Optical Double Resonance (Vol. 37)

Elementary Particle Physics

Current Algebra

Furlan, G., Paver, N., Verzegnassi, C.: Low Energy Theorems and Photo- and Electroproduction Near Threshold by Current Algebra (Vol. 62)

Gatto, R.: Cabibbo Angle and $SU_2 \times SU_2$ Breaking (Vol. 53)

Genz, H.: Local Properties of σ-Terms: A Review (Vol. 61)

Kleinert, H.: Baryon Current Solving SU (3) Charge-Current Algebra (Vol. 49)

Leutwyler, H.: Current Algebra and Lightlike Charges (Vol. 50)

Mendes, R. V., Ne'eman, Y.: Representations of the Local Current Algebra. A Constructional Approach (Vol. 60)

Müller, V. F.: Introduction to the Lagrangian Method (Vol. 50)

Pietschmann, H.: Introduction to the Method of Current Algebra (Vol. 50)

Pilkuhn, H.: Coupling Constants from PCAC (Vol. 55)

Pilkuhn, H.: S-Matrix Formulation of Current Algebra (Vol. 50)

Renner, B.: Current Algebra and Weak Interactions (Vol. 52)

Renner, B.: On the Problem of the Sigma Terms in Meson-Baryon Scattering. Comments on Recent Literature (Vol. 61)

Soloviev, L. D.: Symmetries and Current Algebras for Electromagnetic Interactions (Vol. 46)

Stech, B.: Nonleptonic Decays and Mass Differences of Hadrons (Vol. 50)

Stichel, P.: Current Algebra in the Framework of General Quantum Field Theory (Vol. 50)

Stichel, P.: Current Algebra and Renormalizable Field Theories (Vol. 50)

Stichel, P.: Introduction to Current Algebra (Vol. 50)

Verzegnassi, C.: Low Energy Photo and Electroproduction, Multipole Analysis by Current Algebra Commutators (Vol. 59)

Weinstein, M.: Chiral Symmetry. An Approach to the Study of the Strong Interactions (Vol. 60)

Electromagnetic Interactions

Deep Inelastic Lepton Scattering

Drees, J.: Deep Inelastic Electron-Nucleon Scattering (Vol. 60)

Landshoff, P. V.: Duality in Deep Inelastic Electroproduction (Vol. 62)

Llewellyn Smith, C. H.: Parton Models of Inelastic Lepton Scattering (Vol. 62)

Rittenberg, V.: Scaling in Deep Inelastic Scattering with Fixed Final States (Vol. 62)

Rubinstein, H. R.: Duality for Real and Virtual Photons (Vol. 62)

Rühl, W.: Application of Harmonic Analysis to Inelastic Electron-Proton Scattering (Vol. 57)

Experimental Techniques

Panofsky, W. K. H.: Experimental Techniques (Vol. 39)